HOTHOUSE EARTH

HOTHOUSE
· EARTH ·

The Greenhouse Effect and Gaia

John Gribbin

GROVE WEIDENFELD
New York

Published by Grove Weidenfeld
A division of Wheatland Corporation
841 Broadway
New York, New York 10003-4793

Published in Canada by General Publishing Company, Ltd.

First published in Great Britain in 1990 by Bantam Press, a
division of Transworld Publishers Ltd., London.

Library of Congress Cataloging-in-Publication Data

Gribbin, John R.
Hothouse earth: the greenhouse effect & gaia/John Gribbin. —
1st Grove Weidenfeld ed.
p.   cm.
Includes bibliographical references.
ISBN 0-8021-1374-5 (alk. paper)
1. Greenhouse effect, Atmospheric. 2. Global warming.
3. Atmospheric carbon dioxide — Environmental aspects. I. Title.
QC912.3.G75   1990
363.73'87 — dc20   90-30871
                                                     CIP

Manufactured in the United States of America

Printed on acid-free paper

First American Edition 1990

10 9 8 7 6 5 4 3 2 1

Acknowledgements

Many people gave up their time to talk to me about the greenhouse effect, or provided copies of their publications on the topic (or both) during the preparation of this book. I'd especially like to thank (in no particular order of importance): Jim Hansen, of the NASA Goddard Institute in New York; Mick Kelly, of the Climatic Research Unit in Norwich; Jerry Mahlman, of Princeton University; Paul Crutzen, of the Max Planck Institute for Chemistry; Karin Labitzke, of the Free University of Berlin; Michael Farrell, of the Carbon Dioxide Information and Analysis Program, US Department of Energy; Jim Lovelock, from Coombe Mill; Stephen Schneider, of the US National Center for Atmospheric Research (NCAR); Nick Shackleton, of the University of Cambridge; Ralph Cicerone, of NCAR; Stephanie Pain, of *New Scientist*; and Veerhabadrhan Ramanathan, of the University of Chicago.

I am also grateful to Silke Bernhard and the Dahlem Konferenzen for allowing me to attend two workshops in Berlin where problems related to the changing atmosphere of our planet and the biological productivity of the oceans were discussed. They provided invaluable background material, as well as an opportunity to meet many of the scientists involved in research into the greenhouse effect.

John Gribbin
January 1989

Contents

People sometimes have the attitude that 'Gaia will look after us'. But that's wrong. If the concept means anything at all, Gaia will look after herself. *And the best way for her to do that might well be to get rid of us.*

Jim Lovelock
Berlin, November 1987

Introduction

In 1988, the greenhouse effect moved firmly into the arena of political debate. Carbon dioxide, and other gases, produced by human industrial activity are building up in the atmosphere, and ought, according to theory, to be trapping heat near the ground and making the world warmer. But although scientists had been concerned for years that the world might be destined for a warming caused by human activities, they had been unsure about how big this warming might be, and how quickly it might set in; until the late 1980s, there was very little direct evidence that temperatures world-wide were, in fact, rising. Against this background of uncertainty, even those politicians who were aware of the potential importance of this so-called greenhouse effect for future generations adopted a policy of 'wait and see'. The time to worry about the greenhouse effect would be when there was some real evidence that the world really was getting warmer. That evidence came in with a bang in 1988, causing politicians to sit up and take notice.

One of the key events that shifted the greenhouse debate towards the centre of the political stage was a conference on 'The Changing Atmosphere', held in Toronto at the end of May 1988, and attended by both scientists and policy makers from forty-eight nations, together with representatives of fifteen international agencies and forty-seven non-governmental organizations such as Friends of the Earth. The meeting was hosted by the Canadian government, and supported by the World Meteorological Organization and the United Nations Environment Programme. The picture delegates were presented with was one in which the average temperature of the world will rise by up to 4°C over the next forty years, with sea levels rising as a result by up to 140 centimetres. Centres of population such as the Nile and Ganges deltas, the Yangtse, the Mekong and the Mississippi are at risk, and in one of the more poignant political communications of the year the Prime Minister of Tuvalu, a tiny island nation in the South Pacific, formally asked the United Nations for advice on the problem – the highest point of his country is less than five metres above the waves today.

Delegates to the Toronto meeting were also told that the global warming will proceed unevenly, with high latitudes (nearer the poles) warming more quickly than the equatorial regions. Both drought in Africa and floods in other parts of the world were linked by experts to the burgeoning greenhouse effect, already at work, and the conclusions of the meeting were assured of headline coverage in the media by the United States drought of spring/summer 1988, the worst since the dust-bowl years of the 1930s. That coverage included front page stories in the *New York Times* and the *Guardian*, and leader comment in *The Times*, as the Toronto meeting called for the wealthy industrialized countries to reduce their emissions of carbon dioxide – the gas chiefly responsible for the greenhouse effect – by twenty per cent by the year 2005. The final statement from the Toronto meeting concluded 'humanity is conducting an enormous, unintended, globally pervasive experiment whose ulti- mate consequences could be second only to a global nuclear war', and urged that it is 'imperative to act now'.

This was more than just pious pleading. After years of effort, in 1987 an international agreement aimed at protecting (or, at least, reducing damage to) the global environment had been reached for the first time, through the good offices of UNEP. This Montreal Protocol should lead to cuts in the emissions of chlorofluorocarbons (CFCs), gases that damage the ozone layer of the atmosphere (and also, through a quite separate process, contribute to the greenhouse effect). Environmentally concerned bodies, including UNEP itself, were determined to maintain the momentum, and get to grips with the greenhouse effect.

Against this background, Congressional Committees in Washing- ton and the Environment Committee of the House of Commons in London both heard expert evidence on the changing climate. Experts in both the US and Britain presented similar evidence to their respective elected representatives, and drew similar con- clusions. But it was the American version that hit the headlines on both sides of the Atlantic, partly because such testimony is immediately available to the media in the US, while in Britain it has to await official clearance, and partly because of the way one man, in particular, put the message across in Washington.

James Hansen, of the NASA Goddard Institute of Space Studies in New York City, told the Senate on 23 June 1988 that 'the Earth is warmer in 1988 than at any time in the history of instrumental measurements ... the global warming is now sufficiently large that we can ascribe with a high degree of confidence a cause and effect relationship to the greenhouse effect and ... the greenhouse effect is already large enough to begin to affect the probability of occurrence of extreme events such as summer heat waves ... heatwave/drought occurrences in the Southeast and Midwest United States may be more frequent in the next decade'.

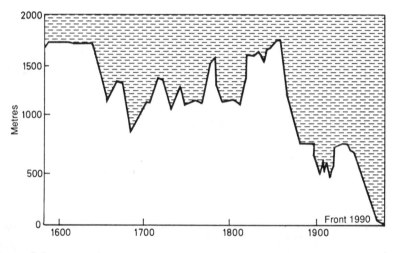

Figure I.1
The change in the climate of the world over the past hundred years or so is dramatically illustrated by the way mountain glaciers have retreated. This graph plots the change in position of the front of the Grindewald glacier, in Switzerland, relative to its present position, over the centuries since 1500. The glacier has retreated by about 1.5 kilometres since 1850, and about half of this retreat has occurred since 1940. The retreat of the glacier mirrors the rise in global temperature (see Figure 1.1, page 9).
(Source: Climatic Research Unit, University of East Anglia)

This was a particularly significant statement – backed up as it was by hard evidence – because Hansen has been studying the greenhouse 'problem' for many years. As well as being an acknowledged world authority on the subject, until 1988 he also had a reputation for caution and was a proponent of the 'wait and

3

see' approach. But by 1988 he had seen enough, and if his formal testimony to the Senate seemed dramatic enough to anyone who knew the background, what he had to say to the press caused a sensation. Acknowledging that it is impossible ever to be a hundred per cent certain that the observed global warming is caused by the greenhouse effect, Hansen stressed that the best statistical tests showed that there was a ninety-nine per cent certainty that the warming is caused by human activities, and said that in the face of that evidence 'it is time to stop waffling so much and say that the evidence is pretty strong that the greenhouse effect is here'.

His words, and those of other experts who testified at the same hearings, did not fall upon stony ground. Senator Timothy Wirth commented 'the scientific evidence is compelling: the global climate is changing as the Earth's atmosphere gets warmer. Now, the Congress must begin to consider how we are going to slow or halt that warming trend and how we are going to cope with the changes that may already be inevitable.' Senator Bennett Johnson said that the hearing 'convinced me that there is a greenhouse effect giving us a global warming'. But the most dramatic short-term political repercussions from Hansen's testimony were felt not in Washington, but when the ripples spread to London.

At the end of September, 1988, experienced observers of the political scene in Britain were taken aback by an unexpected shift in government comments on environmental issues. First, Prime Minister Margaret Thatcher made a speech to the Royal Society in which she drew attention to the importance of environmental issues such as acid rain, the damage being done to the ozone layer, and the greenhouse effect. Then, a few days later, Foreign Minister Sir Geoffrey Howe called on the United Nations General Assembly to hold 'a serious debate' about the threat of climatic change. 'We are,' he said, 'totally dependent on climate. Damage it beyond repair, and the Earth becomes a lifeless desert, spinning in space. We cannot leave a problem of this magnitude to technical bodies.' The words could have come straight from a pamphlet published by Friends of the Earth. They caused astonishment in Britain because the Conservative government, under Margaret Thatcher and with Sir Geoffrey Howe as a senior member throughout the previous nine years, had an abysmal record on environmental issues, doing little to respond to the concern of Britain's neighbours in Europe about acid rain caused by pollution from British power stations, dragging behind the rest of the European Community in efforts to establish controls on emissions of CFCs, and never having acknowledged

4

that climatic change was an issue worthy of serious discussion at home, let alone at the UN.

Diligent reporters soon found that the reason for the change of heart was simple: the Prime Minister had been reading Hansen's Congressional testimony, and had been persuaded by the evidence he presented that, indeed, the time to wait and see had passed. According to the stories that emerged, which doubtless grew a little in the telling, advisers who had for years been trying unsuccessfully to get the Prime Minister to take these issues seriously were astonished when, after reading the testimony, she berated them for their failure to keep her informed on the greenhouse issue. Whether or not any of those advisers had the nerve to point out that all of the evidence was available in the minutes of a House of Commons Environment Committee meeting dating from March 1988, legend does not reveal. And, in truth, it doesn't matter by which route senior politicians have become aware of the greenhouse effect. What does matter is that this global warming is now acknowledged to be real not just by scientific experts such as Hansen, but by politicians who have the authority to take action in response to that threat. Our world is going to change, not just because of the climatic changes resulting from the greenhouse effect, but from the repercussions of political and economic decisions that are made in an attempt to reduce the size of this effect and to adapt to the climatic changes. This book will give you a feel for what is happening to our weather, and outline some of the possible ways in which society, and individuals, might have to respond. Should people who live near the sea start planning to move inland? Is nuclear power now to be regarded as environmentally desirable, because it produces no carbon dioxide? Will hurricanes, like the recordbreaking Gilbert of 1988, become stronger and more common as the tropical seas warm? Will the icecaps melt, and are skiing holidays soon to be a thing of the past? Are droughts in the Sahel region of Africa, Ethiopia, and even the US Midwest now to be a permanent feature of our planet? All these, and other, questions will be addressed in their turn. But the appropriate place to start the story is with the evidence that persuaded James Hansen, the US Senate and the Prime Minister of Great Britain that it is time to cut the waffle – the evidence that the world is rapidly getting warmer.

ONE

The Warming Now

The world *is* getting warmer. The year 1988 was the warmest since accurate records began in the nineteenth century, just beating a record established only in 1987. Overall, the decade of the 1980s is likely to be the warmest ever recorded, with temperatures now roughly half a degree Celsius above those that prevailed a little over a hundred years ago. If the trend continues, we may be in serious trouble in the twenty-first century – and the trend is expected to continue, because the warming so far is closely in line with predictions based on estimates of the greenhouse effect caused by a build-up of carbon dioxide and other gases in the atmosphere as a result of human activities. Nobody worried too much about this as long as it was 'just a theory'; now that the warming is detectable, the forecasts of even more rapid warming in the decades ahead have to be taken seriously. So what is the evidence that the world is now warmer than it has been for at least several hundred years?

It is, in fact, no easy task to find out how much the world has warmed over the past hundred years or so. The records that are available from the earlier part of what climatologists call 'the historical record' are scattered at different sites around the globe; most observing stations were in Europe, and coverage spread out as the decades passed. Something approaching a genuinely global coverage was not achieved until well into the second half of the twentieth century. Even when records are available, it is not easy to interpret them. They were not always made with standardized instruments under standard conditions – for example, at the same time each day – and some of the variations in the long-term trend of temperature at a single observing station may simply be due to changes in the way measurements were made, for example when one observer retired and a new one took his place. I won't bore you with the details, which you can easily imagine for yourself; but as a result of these difficulties, less than a handful of research groups today are recognized as having the expertise needed to determine genuine global and regional trends in temperature from the mixed bag of records ancient and modern. I shall concentrate on just two such studies, one carried out in Britain and the other in the US; the fact that these independent studies, and those of the few other experts competent to assess the records, agree in large measure on the extent of the global warming is confirmation that the effect is real.

Temperature trends

One team, at the University of East Anglia in Norwich, first reported their evidence of a warming trend in both the northern and southern hemispheres of the globe in 1986. This was a key step forward, because the records from both hemispheres showed much the same warming trend, up by about 0.5°C in two stages, first between 1920 and 1940 and then over the years since 1970. The early 1980s were, at that time, the warmest years recorded in both hemispheres, and although the northern hemisphere had cooled slightly in 1984 and 1985, climatologists explained this as due to the effects of the eruption of the volcano El Chichón in South America in 1982. Large, explosive volcanic eruptions send huge amounts of dust into the upper atmosphere, and this acts as a sunshield, blocking out heat that would otherwise reach the surface of the Earth. (An even bigger dip in northern hemisphere temperatures occurred in the 1880s, following the famous eruption of Krakatoa.) In 1986, however, nobody could say whether the cooling of the previous two years was simply a temporary setback in the warming trend, caused by the volcanic dust, or whether the warming trend was over and a new cooling trend might be setting in. These questions were answered within a couple of years.

In February 1988, the East Anglia team updated their global temperature survey, adding in new data from recording stations that had not been included in the previous survey and taking the records forward through to the end of 1987. Adding in the extra recording stations made a slight difference to the figures – the northern hemisphere average for 1981–1984, for example, was reduced by 0.04°C in the revised figures. But these minor adjust-

8

ments paled into insignificance alongside the main conclusions of the study. Averaging over the whole globe, 1987 was now seen as the warmest year since the start of reliable observations in 1858, while 1981 and 1983 tied in second place, just 0.05°C cooler. The downward blip in the record of the mid-1980s in the northern hemisphere was now seen as no more than a temporary lull in the upward trend. The figures for the southern hemisphere were even more dramatic. There, seven of the eight warmest years on record occurred in the 1980s, with 1987 again the warmest. In the northern hemisphere alone, 1987 was just a little cooler than 1981 (perhaps the lingering effect of that volcanic dust). And, remember, 1988 is now breaking those records. The 1980s stand out as something quite remarkable in the historical record of temperature changes (*Figure 1.1*).

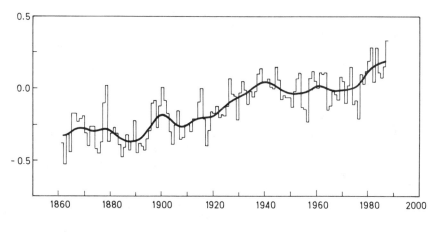

Figure 1.1
The rising trend of global mean temperatures, measured as deviations from the average for 1940 to 1960. In spite of large fluctuations from year to year, the average over successive five-year intervals (smooth curve) shows an increase of about 0.5°C over the past hundred years.
(Source: CRU, University of East Anglia)

These particular measurements all come from observing stations on land. But researchers at the UK Meteorological Office have been studying records of sea surface temperatures measured by observers on board ships since the middle of the nineteenth century. This kind of study is a saga in itself: if it is difficult comparing measurements made by different people using different instruments at different places on land, it is at least doubly difficult working out exactly what has been measured when somebody, a hundred years ago, hauled up a bucket of sea water to the heaving deck of a ship in the middle of the Atlantic or Pacific oceans and stuck a thermometer in it. Many factors affect how much the water warms up, or cools down, while it is on deck and therefore affect the measured temperature. How long was the bucket allowed to stand on the deck before the temperature was taken? Did it stand in sunlight, or in the shade? Was the bucket 'pre-cooled' by giving it a quick dip in the briny before the real sample was taken? Was the bucket made of wood, canvas or metal? And so on. The answers to some of these questions can never, in some cases, be known. Reconstructing past temperature patterns depends on obtaining many different measurements, and using statistical techniques not only to take averages in the right way (balancing, for example, a few readings from a large area of ocean against many measurements from a small land mass) but also to eliminate readings which are shown, by those statistical tests, to be unreliable. Fortunately, the bucket technique was more or less standardized by 1900, so there are relatively few problems in interpreting the all-important record of sea temperatures during the twentieth century, although the experts do disagree slightly in their estimates of sea temperatures back in the 1850s and 1860s. On all interpretations, however, the sea surface measurements also show that 1987 was the warmest year since records began in the 1850s, and they show the same two-stage warming during the twentieth century that was detected from the land-based figures.

The sea surface measurements also go a long way towards removing a nagging doubt that some people had about the way the land-based measurements were made. Many of the observing stations where temperatures have been recorded for a long period of time are, for obvious reasons, situated in or near cities or other urban concentrations. In the twentieth century, towns and cities have got bigger, and both industry and private housing have consumed increasing quantities of energy, with a lot of it escaping as waste heat. It is well known that, as a result, large urban areas are now significantly warmer than the surrounding countryside – as many a commuter could testify. Could it be that the alleged warming trend of global climate was actually due to nothing more than this urban heat island effect, distorting the measurements made by the observers in urban areas? Because the sea surface measurements show the same trend this seems unlikely, but just to make sure Phil Jones and his colleagues at the University of

East Anglia have now identified recording stations contributing to the records they use where this effect might be occurring. They do this, in part, by comparing the records from urban and rural sites that are in roughly the same geographical area to see how the two differ. Out of 2666 recording stations used in the main study, the urban warming effect shows up in just thirty-eight cases, thirty-one of them from the United States. Because of the way the figures are averaged, this probably contributes a rise in temperature of only about 0.01°C to the universal warming observed between 1901 and 1980, and it cannot possibly be contributing a spurious 'signal' bigger than 0.1°C. Since the measured rise in global mean temperatures is about 0.6°C, at least half a degree warming cannot be accounted for by the heat island effect.

Even this long-term figure conceals just how rapidly temperatures are rising now. Global mean temperature increased by about 0.5°C between 1880 and 1940, actually decreased by about 0.2°C between 1940 and 1970 (almost halfway back to where it started), and then increased by about 0.3°C between 1970 and 1980, with a continuing rapid rise in the 1980s. Jones has looked in detail at the twenty-year period from 1967 to 1986. Concentrating on the land surface, where most of us live, he has found a total warming of 0.31°C in the northern hemisphere and 0.23°C in the southern hemisphere. This may come as a surprise to some readers, especially those in northwest Europe. I can imagine them thinking that their own experience doesn't seem to bear this out and that there are no signs, in their part of the globe, that the world is getting warmer. That wouldn't be surprising, because one of the most important discoveries shown up by Jones's study of temperature trends in different geographical regions is that the warming is far from evenly distributed – so far from even, in fact, that a large part of Europe, including Britain, *cooled* by about 0.25°C between the mid-1970s and the mid-1980s (*Figure 1.2*).

This patchiness in the temperature record is not just a feature of Europe, or even of the northern hemisphere. In the southern hemisphere, the present warming is most pronounced over Australia, southern South Africa, the region where the tip of South America almost meets the Antarctic Peninsula, and in the region of Antarctica near Australia. In the northern hemisphere, the warming is strongest over Alaska, northwestern Canada, the Greenland Sea, most of the Soviet Union (but especially Siberia), parts of southern Asia, north Africa and southwestern Europe. There is also, though, the newly identified cooling trend over a large part of Europe,

11

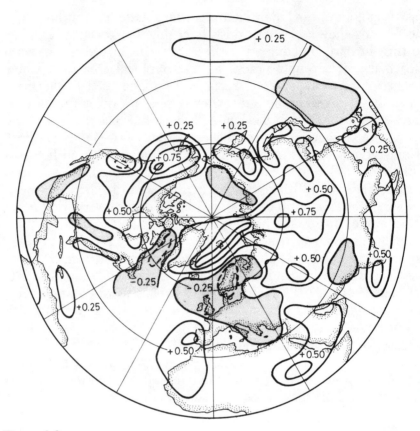

Figure 1.2
Regional temperature trends for recent decades show that the world has not been warming evenly. The numbered contours indicate changes in average annual temperatures in units of tenths of a degree per year. Britain and a large part of Europe have actually cooled during the 1980s; parts of the Soviet Union have warmed dramatically.
(Source: CRU, University of East Anglia)

northeastern and eastern Canada, and southwest Greenland. Over Scandinavia, mean temperatures fell by about 0.6°C at a time when the northern hemisphere warmed by 0.31°C. Scandinavia has got nearly a full degree 'out of step' with its surroundings.

Parts of the Antarctic coast have cooled even more dramatically, by about 1°C, while the strongest recorded warming has reached 2°C over northwestern Canada, 1.6°C over western Siberia, and just over 1.2°C at the tip of South Africa – all over the same twenty-year period.

These figures are based on data from land-based recording

stations, and do not include complete ocean coverage. The anomalous regions of cooling are all on the edges of the oceans, and it may be that there is something going on in the sea that is different in detail from the warming pattern on land. The simplest explanation for the local coolings, however, may be that the oceans have indeed got warmer, and that as a result more water is evaporating from them to form clouds. As the clouds drift over the nearby land, they block out heat from the Sun, causing the land below to cool slightly. If so, then the same parts of the world that are affected by the cooling ought to be experiencing more precipitation – rain or snow depending on their location and the time of year. Raymond Bradley, of the University of Massachusetts, has found that this is indeed the case. The records show that over the past thirty to forty years there has been a significant increase in precipitation at mid-latitudes (including Britain and nearby continental Europe), and corresponding decreases in precipitation at low latitudes (including the parts of northern Africa that now seem to be chronically afflicted by drought and famine). This, as we shall see, very closely matches predictions of the changes that ought to occur in response to a greenhouse effect warming of the world.

An American perspective

James Hansen's team at the NASA Institute in New York has been studying global temperature trends for even longer than the researchers at the University of East Anglia. Like their counterparts in Britain, they have published various studies during the 1980s, updating and improving the information available from the historical record. A key development in this step-by-step progress came in 1981, when the team provided their analysis of all the available data covering the full hundred-year interval from 1880 to 1980.

During the 1970s, climatologists had become used to the idea that the world was in a *cooling* phase, retreating from the high temperatures reached in the early 1940s. Although a few researchers had suggested that the world might soon begin to warm, because of the greenhouse effect, and some calculations had suggested that a very rapid warming might set in before the end of the century, there seemed to be no evidence of this warming actually taking place. But this conception – a misconception, as it turned out – was largely based on records of temperature trends in the temperate latitudes of the northern hemisphere, data from places like Europe and North America, where most of the researchers worked and where they had easy access to long runs of accurate temperature measurements. The NASA team surprised both themselves and a lot of other people when they analysed temperature trends for the entire globe, and found that warming in other parts of the world had, by 1980, already more than compensated for the cooling trend perceived by northerners.

The NASA analysis divided the world up into a grid of eighty boxes, forty in each hemisphere, each with the same area. Temperature trends were calculated for each box, using records from all the observing stations in the box, and then the data for different boxes was combined to give an indication of how temperatures had been changing in different latitude bands, in each hemisphere taken separately and over the globe as a whole. The record begins in 1880, when measurements were being made in at least two widely separated boxes in all of the latitude bands used in the survey except Antarctica; by 1900, there were continuous measurements being made in more than half of the eighty geographical boxes, and the records get better as they come up to date.

Temperature trends from 1880 to 1980 can best be visualized, in this analysis, in terms of three latitude bands. Northern latitudes cover the region from the North Pole to latitude 23.6°N (the Tropic of Cancer), low latitudes are the band from 23.6°N to 23.6°S (the tropical region either side of the equator) and southern latitudes cover everything from 23.6°S (the Tropic of Capricorn) to the South Pole.* The northern latitudes warmed by 0.8°C between the 1880s and 1940, cooled by 0.5°C between 1940 and 1970, and had started to warm again in the 1970s. Low latitudes warmed by about 0.3°C between 1880 and 1930, but then stayed at roughly the same temperature. Southern latitudes warmed more steadily than northern latitudes by about 0.4°C between 1880 and 1980, without the peak in the 1940s and subsequent decline. When all the data is put together to provide a global average, there is still a peak in 1940 following a rise of about 0.5°C since 1880, and a slight cooling afterwards, but by 1980 *global* temperatures were almost

*Because of the way the eighty boxes are arranged geographically, strictly speaking the boundaries between these zones are not precisely at the Tropics, which lie almost exactly at 23.5°C either side of the equator. This doesn't affect the discussion, and it is easier to visualize the division of the globe in terms of the Tropics.

14

back at the 1940 peak (*Figure 1.3*). 'There is,' said the NASA team in 1981, 'a high probability of warming in the 1980s.'

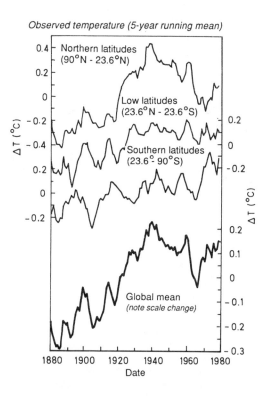

Figure 1.3
Temperature variations in different latitude bands from 1880 to 1980, measured relative to the long-term mean.
(Source: James Hansen/Goddard Institute for Space Studies)

This analysis, published in the journal *Science* in August 1981, may be looked back on as the first recognition that the greenhouse warming had begun. At the time, however, most researchers were cautious about accepting the evidence, and some suggested that there might be something wrong with the figures from the southern hemisphere, which depended on relatively few observing stations, compared with the well-covered north. In a response to criticism of their work, published in *Science* in May 1983, Hansen's team made some comments which stand out today. They mentioned the eruption of El Chichón in 1982, and pointed out that the material blasted into the stratosphere as a result (not only dust, but also tiny droplets of liquid, known as an aerosol) might counteract the warming trend for a year or two, 'but, barring improbable further

15

eruptions . . . significant warming is still likely in this decade.' That forecast, which has been amply borne out by events of the past few years, was based, of course, on calculations of the greenhouse effect, which I shall come on to in a later chapter. While the world was indeed warming as they had forecast, Hansen was keeping track of developments and checking out how the trend grew. By 1987 he was able to look back on the peak warmth of the early 1980s (just before that volcanic eruption), using the same eighty-box analysis to show, with the aid of his colleague Sergej Lebedeff, that this time, unlike the 1940s, the world had warmed evenly, with no sudden surge at high northern latitudes. The broad picture is very similar to the one developed by the University of East Anglia team, and like those researchers Hansen and Lebedeff were back updating their analysis in 1988 to take account of the record-breaking warmth of 1987, which comes out in their analysis, after allowing for urban heat island effects, as 0.63°C above the temperatures of 1880. This time, they pointed out another factor that will have to be taken into account in planning any response to this global shift in climate: whatever it is that is causing the climate to change (and almost certainly the greenhouse effect is at least playing a major part), the climate system today is probably not in a stable state – not in equilibrium. It is being pushed towards a new state, but has not yet settled into it. This could explain regional variations – why, for example, there was rapid warming over Alaska between 1965 and 1985, while southern Greenland cooled – but makes it very difficult to forecast how different regions of the globe will be affected in detail as the rising temperature trend continues.

Whatever they tell us about the scientific side of the problem, though, the set of research papers published by the NASA group in the 1980s shows the pace with which the political side of the story developed. In 1981, it was possible to stand back and take a leisurely look at the record from 1880 to 1980. Comments on the paper, and the NASA team's response to criticism, were published nearly two years later, with no sense of urgency. In 1987, the figures were updated to 1985, chiefly for the neatness of adding another half-decade to the records; those figures were two years out of date by the time they appeared in print. But by early 1988, even one more year's worth of data justified another publication in April, just four months after the last 1987 measurements were made, pointing out the record-breaking warmth now being reached. Even there, Hansen and Lebedeff were cautious about making the connection with the greenhouse effect, merely saying that this was 'a subject beyond the scope of this paper'. But in the four months it had taken to get the 1987 data in print, the world had changed again; just a few *weeks* later Hansen was telling the US Senate that the first five months of 1988 had been warmer than any comparable period since 1880, and that the greenhouse effect was upon us.

Why all the fuss?

Although the greenhouse effect itself, and the implications of the global warming for humankind, are the subject of detailed discussion in the rest of this book, it seems appropriate, in the light of those remarks, to mention briefly why so many people, by the end of 1988, were concerned about the rising trend of temperatures. An increase of less than one degree Celsius in a hundred years doesn't sound very much to get excited about in everyday terms. It is less than the amount by which the tempera-ture varies during the day, or from day to night, and far less than the seasonal variations that inhabitants of temperate latitudes are used to. So why all the fuss?

The decline in rainfall in the subtropical region of the northern hemisphere – the root cause of continuing tragic famine across northern Africa – is the most obvious example to date of why even such small shifts in climate are a cause for concern (*Figure 1.4*). This trend is exactly in line with calculations of how rainfall patterns should change as the world warms, and it has occurred at exactly the time the world has begun to warm rapidly. There is no proof that this is cause and effect, but the circumstantial evidence for a link is strong. In all probability, the greenhouse effect has already killed hundreds of thousands of people. The droughts and associated famines have happened during a rise in temperature of less than half a degree Celsius over the past twenty years, but the forecasts and projections suggest that a rise of a full degree is on the cards over the next twenty to thirty years, with more to follow.

Putting that in perspective, although weather experts are fond of telling you that the climate is always changing, in fact the range of variation that has happened on any human timescale is small. The difference between the cold of the most recent ice age and present-day conditions corresponds only to a global warming of about 3°C. The climatic conditions on Earth today are those of an interglacial, an interval of ten to twenty thousand years between ice ages proper, which themselves last for about a hundred thousand

17

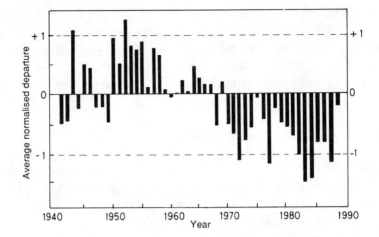

Figure 1.4
Variations in rainfall in the Sahel region of Africa, measured as deviations from the long-term average. The measurements are 'normalized' in such a way that any departure within the range from −1 to +1 might occur by chance, but any departure outside this range is probably a sign of a change in weather patterns. On this evidence, the 'normal' weather in the Sahel is drier now than in the 1950s.
(Adapted from R.A. Kerr, Science *volume 227 page 1453, 1985)*

years. The warmest period of the present interglacial occurred about six thousand years ago. It may be no coincidence that this was the time when humanity began to take steps along the road leading to civilization; but that warmth, the hottest the Earth has been since the ice age, was no more than a degree above the temperatures that prevailed in the 1880s, and half a degree warmer than the 1980s. By the first decade of the twenty-first century, the world will be warmer than it has been at any time since the latest ice age *began*, more than a hundred thousand years ago. By any standards, this is a major human disturbance of the environment.

One way of putting the seemingly small figures in a proper perspective is by thinking in terms of the effect over a whole year, or a season, rather than an individual day. If today is one degree warmer than yesterday, that has no particular significance for the local environment. But if *every* day in a year is one degree warmer than the corresponding day the previous year, there is a big effect. The amount of warmth added to the environment is 365 'degree days', and it is this cumulative effect that melts icecaps in winter (or makes precipitation fall as rain instead of snow), brings droughts in summer and alters the way in which

18

plants grow.* Such a change will also bring with it a change in the frequency with which extreme events occur – the occasional *very* hot, dry spell will become less occasional in the warmer world, and while a reduction in the frequency of sharp winter frosts might seem to be good for crops, it is also good for pests that overwinter in a dormant or larval state. And not just pests that affect plants: in the early part of the twenty-first century, New York, Paris, London and Rome may all be warm enough to suit the mosquito that carries malaria. No matter how small the temperature differences might seem to those of us who live in centrally heated or air-conditioned houses, work in similarly insulated offices, and take skiing holidays in winter but fly off to the sunshine in summer, even the more cautious projections now predict that by the early twenty-first century the world's climate will have shifted into a state that has never existed during the entire time that modern agriculture, on which we all depend for our food, was being developed. Hansen has put this in an American perspective. In the middle part of the twentieth century, the US could expect 'abnormally' hot summers two years in every six; in the 1990s, comparably hot summers will occur, on average, four years out of every six. What was 'abnormal' is becoming the norm; 'normal' summers will be as rare now as 'abnormal' ones were twenty years ago, and the Midwest drought of 1988 is probably only a taste of things to come.

Even if it were 'only' the temperature trend that told us something is happening to change the world's climate, there would be ample cause for concern. But as researchers have become aware of the problem, more and more people have found more and more evidence, from many different lines of study, that we are entering a new climatic era. I have already mentioned the change in rainfall patterns. Several groups have now studied changes in sea level during the past century. This is about as difficult to interpret as the temperature record, and uses similar statistical techniques to unravel the complexities of tide-gauge measurements from around the world, but it has the great merit of being a completely independent test, using a totally

---

*One degree of extra warmth every day, 365 degree days in a year, gives the same amount of extra heat in a year as we would receive in a day if the average temperature rose by 365 degrees. The analogy is not precisely appropriate, but it indicates how much energy we are talking about – a single day with temperature 365 degrees above 'normal' would wipe all life from the surface of the Earth.

different kind of measurement. When allowance is made for seasonal fluctuations and known local effects, the pattern that emerges is one of a rise in global mean sea level of about twelve centimetres in the past century – roughly a millimetre of increase for each year that passes.

Mention rising sea levels, and most people immediately think in terms of melting ice sheets. In fact, this is not the main cause of the increase in sea level since the nineteenth century. When the world gets warmer, the water in the sea will expand, just as a bar of iron expands when it is heated. (If you are making a mug of cocoa, don't measure out how much milk you need by filling the mug to the brim before pouring the milk into the pan to heat – the same measure of milk will be too big for the mug when it is hot.) This thermal expansion of the upper layers of the ocean is enough to account for most of the change that has occurred in sea level this century. Indeed, the rising trend of sea level closely matches up to the changing trend in average global temperatures, even including a flattening off after the 1940s and a renewed rise in the 1970s (*Figure 1.5*). This thermal expansion cannot account for all of the

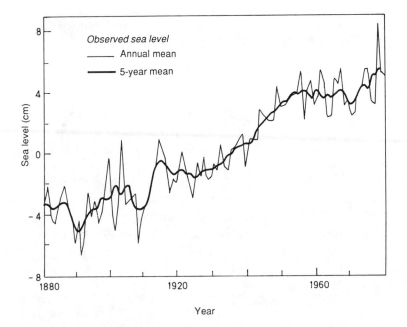

Figure 1.5
The rising trend of global sea level closely follows the rising temperature trend. (Source: James Hansen/GISS)

trend, and new dams and reservoirs built during the past century may have trapped enough water behind them to cause a *fall* in sea level, other things being equal, of one or two centimetres overall; so there probably has been some melting back of glaciers as well. There'll be much more about all this later (especially in Chapter Seven), where I look in detail at the prospects for future increases in sea level; the important point now is that even if we had no measurements of global temperature variations over the past hundred years, the sea level trend itself would be compelling evidence that the world had warmed by about half a degree Celsius. If the trend continues, cities around the Mediterranean will be among those affected most severely in the early part of the next century. Venice is already at risk; much of Alexandria, with a population of 3.5 million, is less than a metre above sea level; the wetlands of the Camargue, in southern France, will be swamped. Whole countries, notably Bangladesh and the Netherlands, are threatened by rising sea levels, as is the state of Florida. It is because the prospects are so grim that scientists have been so cautious about asserting that the greenhouse effect is beginning to bite. But now they have proof.

Another independent piece of evidence pointing the same way emerged in 1986. As a result of the opening up of Alaska to oil exploration, many boreholes were drilled into the frozen ground of this part of the Arctic during the 1970s. In this cold permafrost, heat is transferred only by conduction – there is no liquid water to move from layer to layer, carrying heat with it and confusing the picture. If the temperature at the surface stays the same for a long time, then the pattern of temperature changes going deeper into the permafrost becomes stable, in much the same way that if a bar of iron is kept with one end in a bucket of ice and the other at a constant temperature of 20°C it will establish a stable 'temperature gradient', with the temperature at each point along the bar being a constant and smooth variation from 0°C at one end to 20°C at the other.

But when researchers from the US Geological Survey examined temperatures recorded at different depths down fourteen boreholes covering a region of a hundred thousand square kilometres scattered across northern Alaska, they found that this equilibrium pattern had been destroyed in the upper layers of the permafrost. The pattern of temperature variations that they found can best be explained if a pulse of warmth is slowly penetrating deeper into the permafrost, working its way downward from the surface. The pattern is the same across northern Alaska,

21

and corresponds to a rise in average surface temperatures of about 2°C over the past hundred years. This is higher than the global average, but broadly fits the picture for high northern latitudes inferred by the NASA team (and others) on the basis of historical temperature records; it has the advantage that in the permafrost nature is averaging out all short-term variations (including seasonal fluctuations) and presenting us with a broad picture of the main trends. Although the figures come only from one region of the globe, and are therefore not as convincing on their own as the sea-level variations, they are another completely independent confirmation of the warming.

While some researchers probe down into the ground to determine details of the warming trend, others are looking up into the air. David Karoly, of Monash University in Clayton, Australia, is one of many observers who have been taking the temperature of the atmosphere at different altitudes using instruments carried on unmanned balloons. I mention his work in particular because it covers the southern hemisphere, and helps to show that those critics who doubted the southern figures reported by Hansen's group in 1981 were wrong. Karoly's balloon measurements indicate that in the lower part of the atmosphere (the troposphere, from the surface of the Earth up to an altitude of about ten to fifteen kilometres), the southern hemisphere warmed at a rate of between 0.1 and 0.5°C per decade between 1950 and 1985. The warming became more pronounced between 1966 and 1985, when the data from thirty-one sites around the southern hemisphere indicate warming at rates in the range from 0.2 to 0.8°C per decade.

On its own, this is just icing on the cake. If you haven't already been convinced that the world is getting warmer, then these figures alone won't change your mind. But the balloons – radiosondes – also send back temperature measurements from higher altitudes, and these are particularly significant. Karoly's analysis shows that in the lower stratosphere, the region of the atmosphere immediately above the troposphere, temperatures have been *falling* since the mid-1960s. Once again, the trend is apparent in measurements from around the southern hemisphere, and different observing stations report coolings in the range from 0.2 to 1.0°C per decade. Other researchers have found similar changes in stratospheric temperatures in other parts of the world. Karin Labitzke of the Free University of Berlin has carried out a huge study of northern hemisphere variations, based on data from three thousand observing stations (as ever, there are more observations from the north). This shows an average stratospheric cooling, between latitudes 20°N and 70°N, of 0.34°C between 1966 and 1980.

Why should this cause excitement among the climatologists? Surely if you are looking for a warming trend, and the lower atmosphere is getting hotter but the upper atmosphere is getting colder, it might seem that you only need to take an average in the right way, over the whole atmosphere, for the trend to disappear. But that isn't the case. There is much more air in the troposphere, where the atmosphere is more dense, and the ground and sea are also getting warmer. The stratospheric cooling is nowhere near enough to offset these effects. It is significant for quite another reason: this pattern of changes, with the troposphere getting warmer while the stratosphere gets cooler, is exactly the pattern that is predicted for the greenhouse effect.

There are ways in which the world could get warmer that would not affect the relative balance of heat between the lower atmosphere and the troposphere – if, for example, the Sun got warmer, then we could expect all of the Earth's atmosphere to get warmer as a result. But the greenhouse effect simply rearranges the heat that is available from the Sun. Less heat escapes into space from the warm surface of the Earth because it is trapped by carbon dioxide and other gases in the troposphere. That, of course, makes the troposphere warmer. But because less heat is passed upwards into the stratosphere, the stratosphere cools at the same time that the troposphere warms. The pattern revealed by the radiosonde measurements exactly fits the greenhouse theory, and it is time to explain exactly what the greenhouse effect is, and how it has played an intimate part in the emergence and evolution of life on Earth.

TWO

The Greenhouse and Gaia

If it were not for the greenhouse effect, we would not be here. This may come as a surprise, if you have only heard the words 'greenhouse effect' in the context of the global warming that is now going on and that is perceived as a *threat* to life on Earth.* Strictly speaking, we should distinguish between what might be called the 'natural' greenhouse effect, a result of the presence of a blanket of air around the Earth, and the problem caused by a build-up of trace gases in the air as a result of human activities. The human contribution is usually referred to as the 'anthropogenic' greenhouse effect when we want to make the distinction clear.† But this chapter is chiefly concerned with the natural greenhouse effect, which has been, and remains, a key factor in making our planet a fit home for life.

The natural greenhouse

The Earth is comfortably warm because it is surrounded by a

---

*It will be less of a surprise if you have read my book *The Hole in the Sky*, which is chiefly about the threat to the ozone layer, but touches on the greenhouse effect. If so, I apologize for the inevitable slight overlap of the next few pages with material you have already seen. But don't go away – there is a lot of stuff you haven't seen coming up.

†I'm not suggesting that human beings are not a 'natural' component of the Earth's living systems, and to that extent this choice of terminology might be regarded as imprecise. But what matters is that we all know what is meant by the terms. Like Humpty Dumpty, I believe that it is up to the users of a word to decide what a particular word means.

blanket of air. By 'comfortably warm', I mean that it is a suitable home for life; since liquid water is an essential ingredient of life as we know it, that requires average temperatures on our globe to be in the range where liquid water can exist, between 0°C and 100°C. We don't need to know very much physics to see that the atmosphere of the Earth has something to do with keeping temperatures in this range, since nature has provided us with an alternative example, an airless 'planet' that orbits our Sun, the ultimate source of heat in the Solar System, at almost exactly the same distance that we do. That 'planet' is our Moon. On the airless Moon, the temperature rises to about 100°C on the sunlit surface, and falls to around −150°C at night.

This extreme variation is partly because of the slow rotation of the Moon. Each 'day' there is four of our weeks long. There is ample time for the surface to heat up under the Sun's glare, and just as much time for it to cool during the 'night', when the surface radiates heat away into space. Extreme temperatures are a poor indication of the balance between incoming solar heat and the heat radiated into space in this way. The relevant number is the average temperature of the surface, which is close to −18°C. If the Earth had no blanket of air and was a barren, rocky ball like the Moon, it would also have an average temperature of about −18°C. In fact, the average temperature of our planet in the layer of air just above the surface is about 15°C. The blanket of air keeps the globe some 33°C warmer than it would otherwise be, and allows the existence of oceans of liquid water and living creatures like ourselves.*

How does it do this? In the long run, there must always be an equilibrium between the amount of heat reaching the Earth from the Sun and the amount of heat which the Earth radiates out into space (apart from a relatively tiny amount of heat leaking out from the Earth's core, which I shall ignore). The hotter an object is, the more heat it radiates. If the Earth radiated less heat than it was receiving, it would warm up and therefore radiate more heat. The warming would stop when the outgoing heat was in balance

*The atmosphere also carries heat around the globe, helping to smooth out the temperature differences between day and night, and between the poles and the equator. But that is a quite different process, which does not affect the *average* temperature of the globe.

The numbers here are all approximate, and in round terms you can think of this natural greenhouse effect as keeping the Earth about 35°C warmer than it would otherwise be.

with the incoming heat. If the Earth were radiating more heat than it receives, it would cool until, once again, the outgoing heat balanced the incoming heat. The Sun is hot because nuclear fusion reactions are going on in its core, but the Earth, and other bodies in the Solar System such as the Moon or Mars, are warm only because they intercept some of the solar heat as it passes.

The nature of the energy radiated by a hot object (its wavelength) depends on its temperature. The surface of the Sun is at a temperature of about 6,000°C, and this corresponds to radiation in the visible part of the electromagnetic spectrum, at wavelengths between 0.4 and 0.7 micrometres. It is because the Sun emits radiation at these wavelengths that our eyes have evolved to make use of radiation in this wavelength band, so it is no surprise to find that most of the Sun's energy is radiated as 'visible' light. This radiation passes through the Earth's atmosphere without being absorbed, although some is reflected back into space by clouds. Some is reflected at the surface of the Earth, but a great deal is absorbed and warms the surface of the land or sea.

Most, but not all, of the Sun's energy is radiated in the visible band. About seven per cent of the total energy output of the Sun is at shorter wavelengths, below 0.4 micrometres, known as the ultraviolet; this radiation is absorbed by molecules in the stratosphere, and warms that layer of the atmosphere (*Figure 2.1*). At the other end of the spectrum, above 0.7 micrometres, a little solar energy is radiated in the band known as the infrared. Infrared radiation plays a key role in the greenhouse effect – but not infrared radiation from the Sun.

The surface of the Earth, warmed by solar radiation, has a temperature of a few tens of degrees, Celsius. This is much less than the temperature of the surface of the Sun, so the Earth's surface radiates electromagnetic energy with much longer wavelengths than the radiation from the surface of the Sun. This terrestrial radiation is almost all between four and a hundred micrometres, in the infrared band. Infrared radiation is the heat you feel when you hold your hand near to a warm radiator – indeed, your hand itself, like the rest of your body, radiates infrared heat. It is just like light, but has a longer wavelength. But unlike visible light, radiation in the infrared part of the spectrum is absorbed by water vapour and carbon dioxide in the atmosphere of the Earth. Both these gases absorb strongly in the band from thirteen to a hundred micrometres; water vapour also absorbs strongly in the band from four to seven micrometres. Between seven and thir-

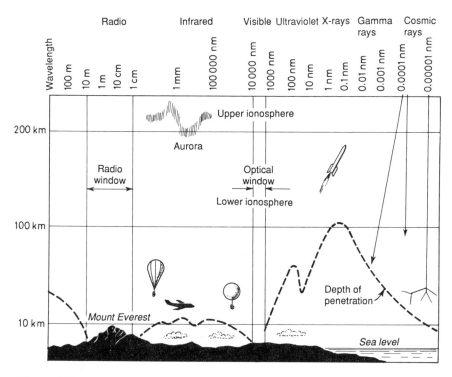

Figure 2.1
Radiation at different wavelengths penetrates the atmosphere of our planet to different depths. Only optical light and some radio waves reach the ground essentially undiminished; ultraviolet radiation is absorbed in the stratosphere, and makes it warm.

teen micrometres there is a 'window' which allows some infrared radiation to escape into space.

The infrared radiation that is absorbed in the lowest layer of the atmosphere, the troposphere, makes the air warm. So the air itself radiates heat in turn, still at infrared wavelengths. Some of this heat goes back down towards the ground, and keeps it warmer than it would otherwise be – the greenhouse effect.*

*This is, in fact, an unfortunate choice of name. A greenhouse keeps the air inside warm in a different way, chiefly because hot air warmed by the Sun is trapped inside the glasshouse and cannot escape by rising upwards in the way warm air usually does – the glass roof physically restrains it. A *little* infrared heat from the ground and plants inside the glasshouse is reflected back inwards by the panes of glass, but this is very much a minor contribution to the warmth inside a greenhouse.

27

The rest works its way upwards through the atmosphere, being absorbed and re-radiated successively until it eventually escapes into space (*Figure 2.2*).

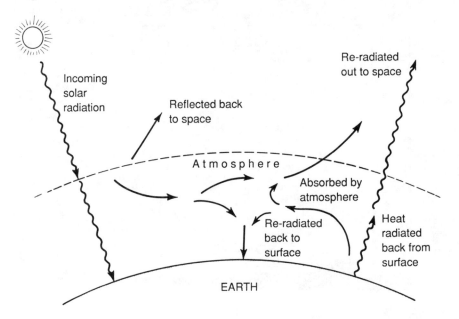

Figure 2.2
The 'natural' greenhouse effect. About forty per cent of incoming solar radiation is reflected back into space, fifteen per cent is absorbed by the atmosphere, and about forty-five per cent reaches the Earth's surface. The warm surface of the Earth radiates at infrared wavelengths, and some of this radiation is absorbed in the atmosphere and re-radiated back to the surface, making it warmer than it would otherwise be.

So the *lowest* part of the troposphere, near the ground, is the warmest, and the temperature of the atmosphere falls off with increasing distance from the surface of the Earth, even though higher layers are closer to the Sun, the ultimate source of heat. This cooling trend reverses in the stratosphere, the layer of the atmosphere above the troposphere. There, temperature *increases* with increasing height, because incoming solar energy in the ultraviolet band *can* be absorbed, by molecules of oxygen and of ozone, the tri-atomic form of oxygen (*Figure 2.3*). But this effect only dominates in the upper regions of the stratosphere; at the bottom of the stratosphere, near the troposphere, heat working its way outward from the surface of the Earth is still an important source of energy. If more heat is trapped near the surface of the Earth, then less heat will be available to warm the upper layers of the troposphere and the bottom of the stratosphere, which is why the simultaneous warming of the lower troposphere and cooling of the lower stratosphere reported by radiosondes is seen as a sign of the anthropogenic greenhouse effect at work, adding to the natural effect.

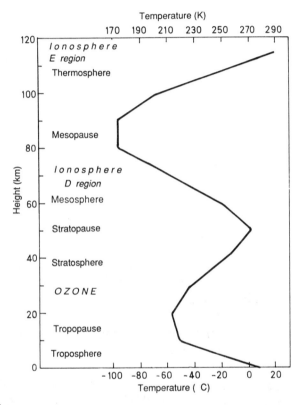

Figure 2.3
The layered structure of the atmosphere of our planet is defined by the way temperature varies with height. Weather systems only circulate in the troposphere, the lowest layer, because the warm stratosphere acts like a lid on convection. The anthropogenic greenhouse effect is making the troposphere warmer and the stratosphere cooler, reducing the effectiveness of this lid.

Although the surface of the globe is about 35°C warmer than it would be without the natural greenhouse effect, the temperature does not keep on rising indefinitely, provided that the composition of the atmosphere stays the same. As long as the amount of water vapour and carbon dioxide (and any other greenhouse gases) is constant, an equilibrium is established, provided the amount of heat coming in from the Sun is constant. Both the ground and the air are warmed by the greenhouse effect, and as the temperature of the atmosphere increases it radiates heat into space more efficiently. For any particular composition of the atmosphere, and any specific amount of solar energy output, there is a balance point, at which the average temperature at the surface of the globe stays much the same. If we were to double the amount of carbon dioxide in the atmosphere overnight, the balance point would shift, and the surface of the Earth would get warmer until the new balance was struck. But while the shift to warmer temperatures was taking place, the atmosphere would not be in equilibrium, and patterns of wind and weather might change in erratic and unpredictable ways. We are not changing the carbon dioxide concentration of the air quite that rapidly. But we have already shifted things away from the natural equilibrium that prevailed a couple of hundred years ago, and the radiation 'budget' of the Earth is not in balance today.

The primeval greenhouse

If it was a surprise to learn that the greenhouse effect is a natural phenomenon that we depend on for our existence, it may also be a surprise to discover that the first person to appreciate this was not some late twentieth-century scientific genius, but a genius from a bygone era, the French mathematician Jean-Baptiste Fourier. Fourier lived in interesting times. Born in 1768, at the time of the French Revolution he was a teacher in the town of Auxerre. He was arrested in 1794, but released a few months later, and then studied in Paris before becoming an assistant lecturer at the Polytechnique. Very much at the centre of things in Paris, he was chosen to accompany Napoleon on his expedition to Egypt in 1798, and later oversaw the publication of the massive *Description de l'Egypte*, a report on the scientific and cultural material brought back by the expedition. He served as a prefect, was made first a baron and then a count by Napoleon, but resigned from government service in protest against the activities of the regime during Napoleon's last days in power. And as if all that weren't enough, he is best remembered for his scientific work, carried out part-time during the Napoleonic years and full-time afterwards, up until his death in 1830 (which was a result of a disease he picked up in Egypt).

Fourier's name is attached to a mathematical series that is widely used to describe repeating (periodic) functions, and has many applications in physics and mathematics. He developed this, though, not as an abstract piece of mathematics, but in an attempt (successful) to find a way to describe how heat is conducted through different materials when they are heated in different ways. He was very much a practical mathematician, particularly fascinated by the nature of heat, and as far as we know he was the first person to suggest, in 1827, that the greenhouse effect keeps the Earth warmer than it would otherwise be. He also seems to have been the first person to use the (incorrect) analogy with the action of the panes of glass in a hothouse, and he even suggested that human activities might modify the natural climate. But it was to be a long time before the full significance of those ideas was appreciated.

The next step forward was taken by a British scientist, John Tyndall (who was actually born in Carlow, now in the Republic of Ireland) in the 1860s. Like Fourier, he was interested in heat, but where Fourier studied conduction, Tyndall concentrated in particular on the transmission of radiant heat through gases and vapours. He actually measured the absorption of infrared radiation by carbon dioxide and water vapour, and published a paper about the effects of water vapour as a greenhouse gas in the *Philosophical Magazine* in 1863. It was titled 'On Radiation Through the Earth's Atmosphere'. Tyndall seems to have been the first person to suggest that ice ages may have occurred during periods when, for some reason, the amount of carbon dioxide in the atmosphere was reduced, weakening the natural greenhouse effect. Among many other contributions to science, he also explained why the sky is blue. (It is because the blue wavelengths of light at the shorter wave – blue – end of the visible spectrum are easily scattered by particles in the air, so that they come at us from all directions, while longer wavelengths, at the red end of the spectrum, are not so easily scattered and therefore come to us more or less straight from the Sun, giving us spectacular sunsets and sunrises.) He also played a part in founding *Nature*, still today one of the world's leading scientific journals, and he died in 1893, only a few years before the next step forward in the greenhouse story.

That step was taken by a Swede, Svante Arrhenius, who is best remembered as a chemist, and who received one of the first Nobel prizes for chemistry in 1903. He was born in 1859, and was active in science long before Tyndall died. But it wasn't until 1896 that he too published a paper on the greenhouse effect in the *Philosophical Magazine*, under the title 'On the Influence of

Carbonic Acid in the Air upon the Temperature of the Ground'. In that paper, he presented calculations of the effect on global mean temperature of a doubling of the natural concentration of carbon dioxide in the air. These calculations were based on measurements of infrared radiation from the Moon at different angles above the horizon, and therefore passing through different thicknesses of air. They had been carried out by an American astronomer, Samuel Pierpoint Langley (who, incidentally, was also an aviation pioneer and was, in 1896, the first person successfully to fly an unmanned powered aircraft; by 1903, at the age of sixty-nine, he had developed a full-size machine capable of carrying a man. He only failed to beat the Wright brothers with that achievement because he tried to launch his aircraft using a catapult technique which proved disastrously unsuitable). The lunar infrared measurements gave Arrhenius a way to work out how much infrared radiation is absorbed in the atmosphere today, and he then calculated that doubling the concentration of carbon dioxide in the air would increase global mean temperature by 5–6°C, not so far from modern estimates, and that the warming would be largest at high altitudes, as modern calculations also suggest. He further calculated that reducing the amount of carbon dioxide to two-thirds of the natural concentration measured in the 1890s would cool the globe by a little over 3°C. In all these calculations, he allowed for the fact that when the world warms more water vapour evaporates from the oceans, and this extra water vapour contributes its own additional greenhouse effect – one of the first scientific calculations to take specific account of what is now known as a 'feedback' mechanism.

The idea of a link between changes in the carbon dioxide content of the atmosphere and changes in climate was taken up by an American geologist, Thomas Chamberlain, right at the end of the nineteenth century. In 1899, he published his paper 'An Attempt to Frame a Working Hypothesis of the Cause of Glacial Periods' in the *Journal of Geology*. His hypothesis involved changes in sea level linked with the growth and retreat of great ice sheets, and associated changes in the carbon dioxide content of the air linked with changes in the rate at which carbon compounds are 'weathered' out of the rocks. Modern climatologists think they have a better explanation of ice age rhythms. But a very similar weathering process was at the forefront of discussions at a meeting held in 1988. That meeting discussed the notion that all the living things on Earth interact, effectively as components of one super-organism, Gaia, that plays a large part in controlling the environment and ensuring, among other things, that our planet never gets too hot or too cold for life to continue. It dealt with, among other things, ice age cycles, something called 'the faint young Sun paradox', the greenhouse effect and the reasons why life exists on Earth but not on Venus or Mars.

Before we move on from the story of the natural greenhouse effect to consider the anthropogenic changes now taking place, it seems appropriate to take stock of just how we got here in the first place, using the evidence presented at that meeting* and beginning at the beginning – with the origin of the Earth and its atmosphere.

Earth, Venus and Mars

If we can learn something about the greenhouse effect by comparing our planet with the airless Moon, we ought to be able to learn a lot more by comparing conditions on Earth with those on other planets that have atmospheres, especially Venus and Mars. These are the two planets whose orbits are closest to that of the Earth: Venus orbits inside our orbit, closer to the Sun, and Mars outside the orbit of the Earth, a little further from the Sun. Together with Mercury, a small, airless planet that orbits even closer to the Sun than Venus does, they form a group known as the 'terrestrial' planets – all relatively small, rocky bodies, quite different from the huge, gaseous planets such as Jupiter and Saturn in the outer part of the Solar System (the 'jovian' planets). Astronomers believe that these neighbours all formed at the same time, in the same way, from a cloud of smaller objects called planetisimals which collided and stuck together. In that cloud of planetisimals, four lumps of rock grew to be bigger than any other aggregations (five including our Moon), and attracted the rest of the pieces by gravity. According to this theory, which is based on a wealth of astronomical evidence including data sent back to Earth by space probes that have visited Venus and Mars, all three of the planets we are interested in here should have started out in much the same state, containing much the same mixture of materials, and even with similar atmospheres. Any differences between them today (and there are many) should simply be a result of their different sizes (Venus is almost the same

*Held in San Diego, under the auspices of the American Geophysical Union, in March 1988.

33

size as the Earth, but Mars is much smaller) and their different distances from the central heat source, the Sun. Today, Venus is a superhot desert, with a thick blanket of air rich in carbon dioxide. Mars is also a desert, but a frozen one, draped with only a thin atmosphere, but one which is also almost entirely carbon dioxide. Earth, of course, is comfortably warm for life, has large oceans of liquid water and an atmosphere made chiefly of nitrogen, with a little less than twenty-five per cent oxygen and a trace of carbon dioxide. The differences are all to do with the greenhouse effect; their origins have been analysed in the most complete detail by three scientists at NASA's Ames Research Center, James Kasting, Owen Toon and James Pollack.*

Toon, in particular, has a long-time interest in dramatic changes to the atmosphere of the Earth. He was a member of the team that first propounded the nuclear winter hypothesis, and he has also made major contributions to understanding the chemical changes in the stratosphere that cause the growth of a 'hole' in the ozone layer over Antarctica each spring. He and his colleagues became interested in the history of the greenhouse effect on Earth and its neighbours from an astronomical perspective, while puzzling over the problem known as the faint young Sun paradox. Astronomers calculate that the Sun must have been as much as thirty per cent cooler than it is today at the time the Solar System formed, a little over 4.5 billion years ago. If the Sun were now to be turned down by this much overnight, the Earth would freeze. Indeed, according to the astronomical calculations, even allowing for the greenhouse effect of the present atmosphere, the Sun would not have been warm enough to thaw the Earth until about two billion years ago. Yet there is clear evidence from sedimentary rocks that there was liquid water on Earth 3.8 billion years ago, and there are fossil traces of life in rocks from 3.5 billion years ago. So how did the Earth stay warm when it was young?

The obvious answer is that the greenhouse effect was stronger then. But what made the effect stronger, and why has it reduced in strength by just the right amount to stop the Earth from overheating, like Venus, as the Sun warmed up? The first major attempt to tackle the 'paradox' was made by Carl Sagan and George Mullen, of

*The NASA team described their work in a very accessible article in the February 1988 issue of *Scientific American*; I go into some of the astronomical background, including the origin of the Solar System and the importance of water for life, in my book *Genesis*.

Cornell University, in the early 1970s. They suggested that there might have been enough ammonia in the atmosphere of the young Earth to keep the planet warm in spite of the Sun being cool. It seemed a reasonable guess, since there is plenty of ammonia in the atmospheres of the jovian planets, so it may well have been a component of the original terrestrial atmosphere. But it turns out that solar energy – even from a Sun thirty per cent weaker than it is today – would split ammonia into its component parts, nitrogen and hydrogen, neither of which is a greenhouse gas. The process is called photodissociation. It may well help to explain where the nitrogen that forms the bulk of the Earth's atmosphere came from; it doesn't explain how the early Earth kept warm.

Several people have suggested that it may simply have been carbon dioxide that did that trick. There is certainly plenty of the right kind of material around today, locked away in rocks in the form of carbonate. All of the atmosphere today exerts a pressure at the surface of the Earth that is defined as one bar – one 'atmosphere' of pressure. If all the carbonate in all the rocks in the Earth's crust were converted into carbon dioxide gas, it would exert a pressure of sixty bars, the equivalent of sixty present-day atmospheres. The amount of carbon dioxide in the air today, making a significant contribution to the 35°C greenhouse effect that keeps the Earth warmer than the Moon, is just 0.035 per cent of the atmosphere, or 0.00035 of a bar. Just one per cent of the amount of carbon dioxide available, half a bar or even less, would be ample to have kept the young Earth warm. Expressed in these terms, it is easy to see that the problem has been turned on its head – it is no longer a puzzle why the young Earth was warm, and instead we have to explain how so much carbon dioxide was taken out of circulation and buried in the rocks quickly enough to keep the Earth cool while the Sun warmed. The key seems to be the presence of liquid water on the surface of the Earth, and one gigantic piece of cosmic luck.

Water is important because it dissolves carbon dioxide. Building on the work of other researchers, the NASA team has proposed that the temperature at the surface of the Earth has stayed roughly constant, never veering to either of the extremes represented by Venus or Mars, because there is a natural negative feedback process at work, which takes carbon dioxide out of the air when the planet gets hotter but puts it back into the air if the planet cools down. It works like this. Carbon dioxide from the atmosphere dissolves in rainwater, forming carbonic acid, which eats away at rocks that contain compounds of calcium, silicon and oxygen (calcium silicates). This chemical action releases calcium and bicarbonate ions, which eventually get into the sea and are used by living organisms, such as plankton, to build their chalky shells, chiefly made of calcium carbonate. When these creatures die, their shells fall to the sea floor, building up layers of sediment rich in carbonate. This thinned the atmosphere at first, but geological activity carries the sea floor (thin 'plates' of the Earth's crust) under the edges of the thicker continental crust that borders the oceans in a process known as 'plate tectonics'; the carbonate is pushed under the continents and deeper into the Earth, where it gets hot and melts. At high temperature and pressure, carbon dioxide is released as new silicate rocks are formed, and the gas finds its way up to the surface and out into the atmosphere during volcanic eruptions.

These geological processes take a very long time to cycle carbon dioxide around. Once the system is established, the amount of carbon dioxide being released into the atmosphere each millennium is roughly constant. So what happens if the temperature of the globe changes? If the temperature falls, less water evaporates from the ocean, so there is less rain, less weathering of the rocks, and less carbon dioxide taken out of the air. As the output from volcanoes continues unabated, the concentration of carbon dioxide increases, warming the Earth through the greenhouse effect and increasing both evaporation and rainfall until a balance is struck. If the world warms, there is more rainfall and more weathering, which takes carbon dioxide out of the air and reduces the greenhouse effect.

The temperature at which the balance is struck depends on the overall tectonic activity of the Earth, and this could be different on other planets, or even on another hypothetical planet orbiting the Sun at the same distance the Earth does. So the carbon dioxide content of the atmosphere of such a twin Earth, and its surface temperature, need not be exactly the same as on our planet even if the two were made from exactly the same mixture of planetisimals to start with. Indeed, temperatures on Earth might vary, over geological time, as a result of changes in tectonic activity, with more active volcanoes in some eras than in others. This might help to explain why, for example, the Earth was lush and warm everywhere during the tens of millions of years that the dinosaurs were around.

The most important point, however, is that this is a *negative* feedback – increasing the temperature as the Sun warms *reduces* the strength of the carbon dioxide greenhouse effect, and vice versa. In the case of water vapour, the opposite is true: increasing temperature puts more water vapour into the air and strengthens the greenhouse effect still further, causing temperature to rise even more, in a positive feedback (as Arrhenius appreciated almost a century ago). That, it seems, is what happened on Venus. The 'wet' greenhouse effect got out of control and became a runaway. And that is where the piece of cosmic luck comes in – our planet is just at a safe distance from the Sun to prevent this happening to us today.

If Earth and Venus indeed formed from the same cloud of planetisimals, then they must have been nearly identical at first, and Venus as well as Earth much have been richly supplied with water. When the Sun was cool, there may even have been oceans on Venus. But the planet soon became hot enough for large amounts of water to evaporate. While the water vapour stayed in the atmosphere, it would add to the greenhouse effect, exacerbating the problem; but it wouldn't stay there for long, because as it filtered up to the higher layers of the atmosphere water molecules would have been ripped apart by sunlight, like ammonia but this time into molecules of hydrogen and oxygen. The hydrogen could then escape into space, and in a few hundred million years all the water would be gone. This process sets in if the incoming solar energy is just ten per cent stronger than at the Earth today; if it is forty per cent stronger, no oceans of water can exist at all on the surface, and the runaway greenhouse develops without going through a wet phase. Today, Venus receives ninety per cent more energy (1.9 times as much) from the Sun as the Earth does. With no oceans and no rainfall, there was no way for Venus to extract carbon dioxide from the atmosphere in large quantities, even if tectonic processes did proceed in the same way there as on Earth; the planet has been left with a ninety-three bar atmosphere almost entirely composed of carbon dioxide, and a greenhouse effect so strong that, combined with the closeness of Venus to the Sun, temperatures at the surface soar to above 500°C. The vital difference between the two planets is that the temperature on Venus passed the critical value needed to trigger a runaway greenhouse very early on in its history.

Kasting, Toon and Pollack calculate that Venus was on the brink of a runaway greenhouse effect as soon as the planet formed. Then, of course, when the Sun was cool, temperatures on Earth were comfortably below this critical threshold. But as the Sun has warmed up, the boundary where the critical temperature is reached has crept steadily outwards across the Solar System. If the present-day orbit of the Earth were just ninety-five per cent of its actual size, then according to the NASA team it would take only a few hundred million years for the planet to lose all its water by photodissociation, and for the runaway greenhouse to take a grip. Since the Sun is still getting warmer, by about one per cent every hundred million years, our planet could be in serious trouble in a billion years' time. If only Mars had been a little bigger it would have had scope for a much longer history as a suitable home for life than our planet, barring some kind of super-engineering by our descendants, is destined for.

We can only speculate that there may once have been oceans of water on Venus. Perhaps the planet was always too hot for this, and it never went through the wet greenhouse phase but straight on to the dry, runaway conditions we see today. On Mars, though, we can actually see traces carved by long-gone rivers. There is no doubt that the red planet was once warm enough for liquid water to flow, and again the natural explanation is that it was kept that warm when it was young, even though the Sun was fainter then and Mars is further from it than we are, by the carbon dioxide greenhouse effect. Calculations show that no more than five bars of carbon dioxide, and perhaps as little as one bar, would have been enough. There was no problem supplying the gas in the beginning if, like Venus and Earth, Mars formed the primeval cloud of planetisimals orbiting the Sun. It *must* have had proportionately as much carbon dioxide and water as the other two planets. So where did it go? Kasting and his colleagues think that Mars originally had a carbon dioxide cycling system, but that this worked in a different way from the one on Earth. There is no sign of plate tectonic activity on Mars, the process which recycles carbon dioxide on Earth and stops it all getting locked away in the rocks. But it is possible that lava flows from volcanoes could have buried carbonate sediments and squeezed them to depths where the heat and pressure would do the same job. Such a process would only work when the planet was young and volcanically active; because Mars is smaller than the Earth, it lost its internal heat relatively quickly, and this early volcanic activity died away less than a billion years after the planet formed. Once that happened, as long as there was still rainfall, carbon dioxide continued to be removed from the atmosphere, but it wasn't replaced. So the planet cooled and eventually froze as the greenhouse effect weakened. According to these ideas –

and other variations on the theme – there must be huge quantities of carbonate rocks buried in the crust of Mars; future expeditions to the red planet will be able to test the hypothesis by looking for these sedimentary rocks.

One implication is that a planet as big as the Earth, in an orbit like that of Mars, would maintain comfortable conditions for life on its surface through the greenhouse effect and its associated negative feedback processes. Indeed, it would provide such a home for life for *longer* than the Earth will, if nature has its way. The fact that the negative feedback process has removed so much of the carbon dioxide from the atmosphere of the Earth today, leaving just 0.00035 of a bar to contribute a relatively tiny greenhouse effect, shows how close we are to overheating. In a billion years from now, when the Sun is ten per cent warmer, there will not be enough leeway in the system to keep the Earth cool. With all the carbon dioxide removed from the atmosphere, the temperature will still rise, putting more water vapour into the air and initiating the fatal runaway wet greenhouse effect. It has taken life on Earth about four billion years to evolve intelligence and notice these interesting features of the greenhouse effect; if it had taken *five* billion years, life on Earth would have been extinct before anyone was around to notice what was going on.

In the orbit of Mars, Earth could escape this fate for far longer. But it wouldn't do to be much further from the Sun than Mars is, or life could never have got started even on an Earth-like planet. Not far beyond the orbit of Mars, the Sun is too faint for the greenhouse effect to have provided a comfortable home for life, even on a planet as big as the Earth, during the history of the Solar System to date. What astronomers call the 'habitable zone' in our Solar System stretches from just inside the orbit of the Earth to just outside the orbit of Mars (*Figure 2.4*). Two planets out of the Solar System's family of nine orbit in that zone – a reasonably large proportion, which looks even more reasonable if you think, as many astronomers do, that Pluto is not a proper planet and increase the ratio to two out of eight. The implication is that other planets circling other stars may also lie in the equivalent habitable zones, and bear life. That is another story; but life, it has been suggested, may play a key role in *maintaining* habitable conditions here on Earth. And Mars plays a part in that story, too.

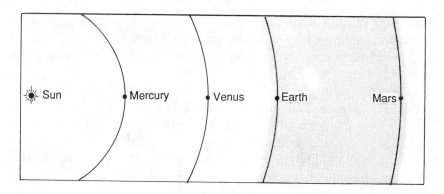

Figure 2.4
The zone of life around our sun embraces the orbits of both Earth and Mars, but not Venus.

Gaia and the greenhouse

I have already mentioned that life on Earth is involved in processing carbon dioxide from the air into the rocks and back again, through the carbonate-silicate cycle. On Earth, carbonate rocks are made from the shells of millions of tiny sea creatures. This doesn't necessarily mean that life is an essential part of the cycle, though. The NASA team, for example, argue that if there were no shelled creatures around, the concentration of calcium and bicarbonate in the sea would simply rise until it reached a critical level where calcium carbonate forms directly – a very simple bit of chemistry that can be reproduced in a school lab. This, they say, is the way the process operated until about six hundred million years ago, when the first shell-makers evolved – 'the Earth would still have remained habitable even if it had never been inhabited' – but others disagree, and see a much more central role for life in the evolution of the planet itself. This brings us right back to that conference in San Diego early in 1988.

Jim Lovelock, the founder of the Gaia hypothesis, is no ordinary scientist. Instead of working in a university department, or some large research laboratory of an industrial corporation, he works at home in the Cornish countryside. Instead of pleasing some head of department or corporate manager, most of the time he does what he likes. The only snag with the life of an independent scientist is the need to make a living for himself and his family. Lovelock likens his way of making a living with that of the artist – perhaps a novelist – who finds no market for the work he really wants to do and turns out commercially acceptable potboilers from time to time in order to survive. Lovelock's 'potboilers' are inventions, and it just happens that he has a genius for inventing sensitive pieces of chemical machinery which can be patented and bring in a useful income. One of the items he developed in the 1950s, for example, was a chemical sniffer, known as an electron capture detector, that can detect minute traces of chemicals in the air. Such instruments revealed the presence of DDT and other pesticides throughout the 'natural' environment, the information that inspired Rachel Carson to write her landmark book *Silent Spring*; the same family of instruments is used to detect explosives in baggage at airports, and to trace the spread of chlorofluorocarbons, the gases that threaten the ozone layer, through the atmosphere of our planet.

In the 1960s, Lovelock spent part of his time working as a consultant with NASA, designing instruments that flew on unmanned spacecraft to probe the chemistry of the Moon and Mars. Commuting between England and the Jet Propulsion Laboratory in California in his role as a consultant, Lovelock had plenty of time to think about the work NASA was doing, and the philosophy behind it. He marvelled at the superb engineering and physical science skills that went into the design and construction of probes such as the Viking landers, but he despaired at the primitive quality of the life sciences part of the programme. In an article in *Co-evolution Quarterly* (Spring 1980) he later wrote:

> It was almost as if a group of the finest engineers were asked to design an automatic roving vehicle which could cross the Sahara Desert. When they had done this they were required to design an automatic fishing rod and line to mount on the vehicle to catch the fish that swam among the sand dunes.

So, Lovelock asked himself, was there a better way to search for traces of life on Mars? A little thought convinced him that observations made from the Earth had *already* shown that there was no life on Mars, a conclusion that did not exactly endear him to his colleagues at JPL who were committed to a multi-million dollar effort to answer the question Lovelock claimed to have answered already. But today, after the Viking craft (equipped with Lovelock's detectors, among other things, but no fishing rods) have failed to find any interesting martian biology, his view is very widely accepted.

It depends upon the fundamentals of chemical systems. Any mixture of chemicals tends to react to produce the most stable molecules possible. Ultimately, it is the laws of quantum physics that determine which molecules are stable and which molecules and atoms react violently – but I don't need to go into those details here.* A mixture of hydrogen and oxygen gases, for example, will explode violently if a spark passes through it, producing stable molecules of water; carbon and oxygen burn to produce carbon dioxide; and so on. I have deliberately chosen two examples that involve oxygen, because oxygen is not only a highly reactive gas, it is also a major component, nearly a quarter, of the atmosphere of the Earth. If our planet were in chemical equilibrium, all this oxygen would be locked away in compounds such as carbon dioxide and water, iron oxides and silicate rocks. 'The gases in the Earth's atmosphere,' says Lovelock in his book *The Ages of Gaia*, 'are in a persistent state of disequilibrium.' They are maintained in that state by the action of life. A world that is in chemical equilibrium must, says Lovelock, be dead; it is a feature of life that it distorts the natural equilibrium. Life is an improbable chemical arrangement, and the influences it produces on its surroundings, through its inhalations and exhalations, are also improbable chemical arrangements.

A hypothetical visitor from another star, approaching the Solar System from outside, could easily analyse the compositions of the atmospheres of the planets, using the same techniques of spectroscopy that we use to study the atmospheres of Mars and Venus from afar. Such a visitor would immediately spot the Earth was the odd one out of the terrestrial planets. Our home has a disequilibrium atmosphere rich in oxygen; Mars and Venus have stable carbon dioxide atmospheres in chemical equilibrium. They are dead.

Having established to his own satisfaction that Mars was dead,

---

*It also has to do with the second law of thermodynamics, the rule that says entropy always increases, or in everyday terms 'things wear out'. Living things, at least for a time, do not 'wear out' but build complexity out of the simple chemicals in their surroundings. They achieve this with the aid of energy from the Sun, and the Sun *is* wearing out, so the little bubble of negative entropy represented by life on Earth, embedded as it is in a bigger bubble of increasing entropy, is not really violating the laws of physics. Even so, it is a sure sign that something odd – life – is going on on our planet.

however, Lovelock continued to muse about the nature of life on Earth, and how it maintained the characteristic chemical disequilibrium. Together with Lynn Margulis, of Boston University, Lovelock developed these musings into a scientific hypothesis, proposing that the living systems of the Earth can be regarded as a single unit which regulates the environment in such a way as to maintain conditions suitable for life. The name 'Gaia', which the Greeks used for the goddess of the Earth, was suggested by the novelist William Golding, who lived in the same village as Lovelock. Gaia does not control her environment consciously, any more than you or I make a conscious decision to start shivering when we are cold. But natural feedback mechanisms involving life, the argument runs, maintain the conditions that life requires automatically. The Earth has stayed hospitable for life for well over three billion years, in spite of many conceivable disasters (such as the changing heat output of the Sun) that could have altered this situation. 'For this to happen by chance,' Lovelock said on the *Planet Earth* TV series in 1986, 'is as unlikely as to survive unscathed a drive blindfold through rush-hour traffic.'

After many years of being regarded as something of a fringe idea, even cranky, in the late 1980s Gaia began to be seen as at the very least providing a framework for discussion of problems involving the living Earth – what Lovelock calls 'geophysiology' – and perhaps as hinting at a deep truth. The San Diego conference was a landmark, with Gaia at last achieving scientific respectability, and Lovelock's second Gaia book describes the way he believes that life has moulded conditions on Earth throughout geological history. In this book, of course, we are primarily interested in just one aspect of the story, but it is one of the most important aspects. How did the Earth maintain an almost constant temperature over the past four billion years?

The researchers who have propounded the carbonate-silicate cycle (originally James Walker and Paul Hays of the University of Michigan, together with Kasting) do not see the role of life as essential. But Lovelock does. Not necessarily the role of the plankton that take carbon dioxide out of circulation and deposit it as carbonate rocks at the floor of the sea, but the role of plants on land that are intimately involved in removing carbon dioxide from the air and boosting the concentration of dissolved carbon dioxide (carbonic acid) in the soil, where it carries out the crucial weathering of silicate rocks:

Consider a tree. In its lifetime it deposits tons of carbon gathered from the air into its roots, some carbon dioxide escapes by root respiration during its lifetime, and when the tree dies the carbon of the roots is oxidized by consumers, releasing carbon dioxide deep in the soil. In one way or another living organisms on the land are engaged in pumping carbon dioxide from the air into the ground. There it comes into contact with, and reacts with, the calcium silicate of the rocks to form calcium carbonate and silicic acid. (*The Ages of Gaia*, page 134.)

According to Lovelock's calculations, it is only the efficiency of this biological contribution to the cycle that makes it possible for the cycle to persist at all with as little as 0.035 per cent carbon dioxide in the air. To sustain the same amount of carbon dioxide in the soil without biological intervention, the amount of carbon dioxide in the atmosphere would have to be about three per cent, a hundred times the present day concentration, and as we have seen that would pose a grave danger of triggering a runaway wet greenhouse.

Lovelock also suggests that life has contributed to maintaining the stable average temperature of the globe in other ways throughout geological history. A little after life on Earth began, he says, there was a sharp fall in temperature, from around 28°C to 15°C. This was due to the decline in carbon dioxide concentrations as the first photosynthesizers (bacteria) used the gas as a source of carbon to build their bodies; but the temperature did not fall any further at that time, a little over 3.5 billion years ago, because those same photosynthesizers released methane, another greenhouse gas. This established a biological feedback. When Lovelock constructed a simple computer model to simulate such conditions, he allowed bacteria to grow at maximum rate when the temperature was 25°C, but with less activity at higher or lower temperatures, and none at all below 0°C and above 50°C. He added carbon dioxide to the atmosphere in line with geological estimates of the productivity of volcanoes, and removed it by weathering at a rate which allowed for the steady development of the ecosystem. Cloud cover in the model is assumed to increase slightly as the land surface of the Earth is colonized by life and plants produce emissions that, as today, provide the 'seeds' on which cloud droplets can form – and, of course, the increasing heat of the Sun was allowed for.

The role of bacterial life is crucial in stabilizing the temperature of this model. When the bacteria are more active, more carbon dioxide is taken out of the atmosphere and more methane is released. When they are less active, there is more carbon dioxide and less methane around. Because the two gases have different greenhouse properties, and are present in the atmosphere in different proportions, changes in biological activity alter the strength of the greenhouse effect. In Lovelock's computer simulations, negative feedbacks ensure that the temperature stays almost constant for more than a billion years, increasing only slightly as the Sun warms. This is as long as it is appropriate to run the model for, because something happened to the Earth about 2.3 billion years ago to produce another temporary drop in temperature; Lovelock identifies this with the emergence of forms of life that release oxygen to the atmosphere. Oxygen would have reacted with methane to remove this component from the greenhouse system; the focus then shifts to other greenhouse mechanisms (discussed in Lovelock's book), but not without an intermediate hiccup in the form of a cold epoch.

Although the details may never be proven accurate, it is quite clear from examples like this that some biological processes influencing the environment must have operated during the history of the Earth, just as the changing environment influenced the evolution of life – what climatologist Stephen Schneider calls the 'co-evolution' of climate and life. The debate about Gaia today focuses not on the question of whether life *can* have influenced the climate (and other aspects of the physical environment) but on whether conditions on Earth have really been *optimal* for life. The subtleties of this argument are beyond the scope of the present book, and I confess that I find my own current opinion on the matter depends on whether the last person I discussed it with was Jim Lovelock or Steve Schneider. But it certainly seems to be established that what we might call the 'weak' version of the Gaia theory holds: that life is not only affected by the physical environment, but affects the physical environment, and often in such a way that conditions suitable for life (though not necessarily optimal conditions) are maintained. The carbon dioxide cycle (or cycles) is one of the most important aspects of this control over the environment by living things, and by taking fossil fuel out of the ground and burning it, as well as by cutting down tropical forests, we are upsetting the natural balance of a key process involved in regulating the temperature of the Earth. Extending the Gaian analogy, we can be likened to a virus infecting Gaia and causing her temperature to rise. Like a body fighting infection, Gaia has several natural responses to this problem, and some of these biological responses will emerge during the discussion of the rest of this book. The greenhouse effect is *not* simply a physical phenomenon. But the fact that Gaia's control over her environment is imperfect, at least on the shorter geological timescales, is amply shown by a close look at the way the Earth's climate has fluctuated over the past few million years. The greenhouse effect is certainly involved; but is Gaia?

THREE

The Ice Age/Carbon Dioxide Connection

The biggest natural changes in climate that have occurred since the emergence of the human species, *Homo sapiens*, half a million years ago have been the switches into and out of ice ages. The background story of the way in which something – possibly Gaia – has maintained temperatures suitable for life on the surface of our planet for several *billion* years is intriguing, but this still allows for quite a wide range of temperatures by human standards. A hundred million years ago, for example, when dinosaurs roamed the Earth, the climate was warmer everywhere than it is today, there were no permanent polar icecaps, and 'tropical' vegetation grew at the latitudes of North America and Europe.

Such conditions, in which dinosaurs flourished, and modern conditions, in which human civilization has arisen, are both within the range that Gaia finds comfortable – just as you or I may be equally happy indulging in winter sports in the mountains or relaxing by the sea in summer. The causes of climatic variations within this range are probably linked to changes in the activity of volcanoes (which alter both the amount of dust high in the atmosphere shielding the Earth from the Sun's warmth and the amount of carbon dioxide present in the air to contribute to the natural greenhouse effect), and especially to changes in the geography of the globe, as continental masses shift their positions and alter the flow of warm and cold ocean currents. But the continents have not shifted very much in the past million years, during which the natural state of the Earth has been slightly colder than today, full ice age conditions with extensive icecaps in both northern and southern hemispheres. Many pieces of geological evidence show that there has been a regular rhythm of temperature changes since before *Homo sapiens* appeared. In this rhythm, an ice age roughly a hundred thousand years long is succeeded by a slightly warmer interval, called an interglacial, which lasts for about ten to twenty thousand years. The pattern has repeated ten times in just over a million years (and was only slightly different in the two million years before that). Today, we are living near the end of a natural warm spell, an interglacial which began a little over ten thousand years ago. These rhythms are now very well understood. Studies carried out in the second half of the 1980s have established beyond doubt that the ice age rhythms are linked with changes in the orientation of the Earth as it orbits around the Sun, and are amplified by changes in the carbon dioxide content of the atmosphere. The *natural* greenhouse effect seems to have an adjustable thermostat, whose temperature setting is partly determined by astronomical influences. The discovery is important because it shows how changes in carbon dioxide concentration of the atmosphere really do affect the temperature of the globe; it is alarming because the build-up of carbon dioxide caused by human activities is bucking against the natural trend, throwing a spanner in the works which may cause unpredictable changes in the workings of the weather machine.

Ice age rhythms

The astronomical theory of ice ages goes back to the nineteenth century, and the work of a Scot, James Croll, who was born in 1821 and published his ideas in the 1860s, almost exactly at the same time that John Tyndall was publishing his work on the greenhouse effect. Croll's first ice age paper was even published in the *Philosophical Magazine*, the same journal that published Tyndall's paper 'On Radiation Through the Earth's Atmosphere', but just one year later, in 1864. At the time, the two theories of ice ages seemed to be rivals; it was 120 years before researchers realized that the two effects discussed by the nineteenth-century pioneers, working together, are needed to explain the ice age rhythms. By that time Croll's name, although not by any means forgotten, had been relegated to the footnotes of history in most accounts of the astronomical theory, which is today often called the 'Milankovitch Model', after a Yugoslav astronomer, Milutin Milankovitch, who refined and improved the astronomical theory in the decades before the Second World War.

48

Whatever name it goes by, the cycles that are the basis of the astronomical theory are easy to understand. The difficulty lies in explaining how the modest seasonal changes in temperature they produce can cause the ice age/interglacial pattern – which, as we shall see, is where carbon dioxide comes in.

We experience seasonal changes in weather because the Earth is tilted relative to the plane of its orbit about the Sun. Imagine a line drawn from the centre of the Sun to the centre of the Earth. This line lies in a plane that astronomers call the plane of the ecliptic. If our planet were upright, it would be spinning in such a way that the North Pole lay directly above this line, so that a line from the pole to the centre of the Earth would meet the first line at right angles. If that were the case, everywhere on Earth there would be twelve hours of daylight and twelve hours of night every day of the year. There would be no seasonal variations. In fact, the spinning Earth leans over slightly 23½ degrees out of the vertical – 23½ degrees away from the perpendicular to the ecliptic (*Figure 3.1*).

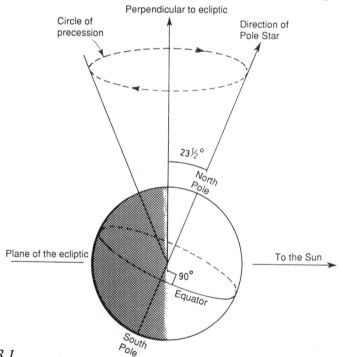

Figure 3.1
The Earth is tilted by about 23½° out of the perpendicular to a line joining the Earth to the Sun. Changes in the angle of tilt, and in the way the tilted Earth wobbles, or precesses, over thousands of years explain many features of climatic change and ice age cycles.

49

Because the spinning Earth acts like a gyroscope, it maintains this orientation very accurately, even though it is moving around the Sun. In the course of a year, although the Earth's position in its orbit around the Sun changes, the direction in which it is tilted stays the same, with the North Pole always 'pointing to' the Pole Star. This is what causes the seasons.

When the Sun is on the same side of the Earth as the Pole Star is, the northern hemisphere is tilted towards the Sun. There are twenty-four hours of sunlight at the pole, and more sunlight hours than dark each day over the whole northern hemisphere, where it is summer (*Figure 3.2*). Six months later, the Earth is on the other side of the Sun, but still tilted towards the Pole Star. Now, the northern hemisphere is pointing away from the Sun. The pole is dark, and there are long nights everywhere in the northern hemisphere, where it is winter. The seasons are reversed in the southern hemisphere, and the two intermediate seasons correspond to intermediate positions of the Earth in its orbit around the Sun.

Figure 3.2
The tilt of the Earth also causes seasonal variations.

All this happens because the tilt of the Earth is constant over a year. But as I said, the spinning Earth is like a gyroscope, and anyone who has watched a child's spinning top will know that a gyroscope can wobble as it spins. The Earth wobbles in a similar fashion, chiefly as a result of the gravitational forces tugging on it from the Sun and Moon, and also because of the gravitational influence of the other planets. But because the Earth is such a big 'spinning top', the wobbles are rather slow. The wobbles are so slow that they don't affect the seasonal patterns over a year, or even a hundred years; but they begin to become important when we look at stretches of time measured in thousands of years. The most rapid change we have to worry about is that the direction in which the North Pole points actually traces a circle on the sky, following a cycle which varies slightly in length but has two main components, near twenty-three and nineteen thousand years. Ten thousand years ago, the present day Pole Star would *not* have provided an accurate indication of the direction north. The shifting orientation of the pole causes an apparent drift of the stars across the sky, as viewed from Earth, which is called the precession of the equinoxes; in fact, it is the spinning Earth that is precessing, and one result of this is that the pattern of the seasons changes slowly as the millennia pass. At the same time, the angle at which our planet is tilted varies, nodding up and down between 21.8° (more nearly upright) and 24.4° (more tilted) over a cycle roughly forty-one thousand years long. The present day tilt, which is actually closer to 23.4° than 23.5°, is about halfway between these extremes. At present the tilt is decreasing, which means, among other things, that the differences between summer and winter are less today than they were a few thousand years ago. Summers are a little cooler and winters a little warmer than they used to be.

The third component of the Milankovitch Model is a slightly different effect, linked with changes in the shape of the Earth's orbit around the Sun. This is also caused by the shifting interplay of gravitational forces in the Solar System; as a result, the Earth's orbit stretches slightly from circular to elliptical and back again with a period of roughly a hundred thousand years. When the orbit is circular, the Earth receives the same amount of heat from the Sun every day of the year; when the orbit is more elliptical, on some days of the year our planet is closer to the Sun, and receives more heat, than on other days in the same year. But the *total* amount of heat received by the whole planet over a whole year is always the same (as long as the Sun's output is constant); the three astronomical cycles *only* redistribute heat between the seasons, and do that by a very modest amount.

At the time of Croll, or even of Milankovitch, it was impossible to test the astronomical theory of ice ages properly, because nobody knew the exact dates of the advances and retreats of the ice over past millennia. Most climatologists regarded the theory as extremely implausible, because the astronomical calculations show that the change in seasonal heating from the Sun – or 'insolation' – is small. In one of his earliest calculations, Milankovitch himself had presented the figures in terms of the way summer sunlight has varied at latitude 65°N. Compared with the present day insolation at different latitudes, the summer insolation at 65°N has varied, over the past six hundred thousand years, between the equivalent of present day summer insolation at 70°N and present day summer insolation at 60°N. But, of course, in the years when summers were hotter, winters were colder (and vice versa). The effects exactly balance, so that, like all latitudes, the band around the globe at 65°N always received the same amount of insolation over the whole year as it does today. The Milankovitch Model remained an intriguing but unproven hypothesis, with only a few ardent supporters, until techniques were developed to trace the temperature of the globe from millennium to millennium down through the ages. The breakthrough came in the 1970s, from studies of the remains of the chalky shells of planktonic foraminifera in deep sea sediments.

These sediments are constantly being laid down as plankton die and fall to the sea bed. Older remains lie deeper below the surface and younger sediments near the top. The sediments can be extracted from the sea bed in the form of long cores drilled by geological exploration ships, but the age of the sediments at any particular depth cannot be simply read off from the core; there is no straightforward equivalent to the annual growth rings of trees, which can just be counted to find the age of a particular piece of wood. Instead, the sediments have to be dated by comparing the magnetism locked up in the sediments with the magnetism of rocks from the continents that have already been dated by other means.

The Earth's magnetic field varies considerably over geological time, sometimes being weaker, sometimes stronger, and sometimes reversing direction completely. This pattern of changes, especially the reversals, provides a distinctive fingerprint. A stretch of sediments that shows a particular magnetic fingerprint can be matched up against its continental counterpart, and dated very accurately. So the geologists knew that a particular layer of deep sea sediment had been laid down between, say, thirty-five and thirty-six thousand years ago, that another layer was deposited between 127,000 and 128,000 years ago, and so on. In fact, by the mid-1970s they had a continuous record of sediments covering some half a million years back from the present day, extracted from the southern ocean. But this only gave them half the story. The other half of the problem was using this sediment core as a fossil thermometer, tracing the temperature at the time each layer was being laid down.

The trick depended on some subtle measurements, and an understanding of how the biochemistry of foraminifera is affected by changing global temperatures. There are two common types of oxygen atoms (two isotopes) known as oxygen-16 and oxygen-18. Both are present in the air that we breathe, and also in sea water (H_2O); since oxygen-18 is heavier than oxygen-16 this means that some molecules of sea water are heavier than others. Heavier molecules of water freeze more readily than lighter molecules, so during an ice age proportionately more of the heavy molecules are locked away in ice. As sea creatures take oxygen from their surroundings to help build their shells, the shells they build during an ice age contain proportionately more of the *lighter* oxygen isotope, which has not been locked away in the ice sheets. By measuring the way the proportions of the two oxygen isotopes vary in different layers of sediments, it is possible to infer not the exact temperature at the time the sediments were laid down, but the extent of global ice cover. The sediments provide a direct record of the pulsebeat of ice ages.

Finding the right sediments to analyse, getting the timescale right, and extracting the isotope data were all difficult tasks. Various aspects of the problem were tackled by various researchers from the 1940s onwards. The key breakthrough, however, came in 1976, when Jim Hays of the Lamont-Doherty Geological Observatory, John Imbrie from Brown University, and Nick Shackleton of the University of Cambridge put together an analysis of the changing pattern of ice cover revealed from the isotopes in core sediments spanning half a million years. The analysis showed that the rhythms of advance and retreat of ice cover over the past few hundred thousand years are dominated by three cycles, respectively a hundred thousand, forty-one thousand, and a little over twenty-one thousand years long. The Milankovitch Model had well and truly come in from the cold. But the puzzle remained: how could such small, seasonal effects produce such profound changes in global climate?

Strengthening the link

In the ten years following the work by Hays, Imbrie and Shackleton, the link between ice ages and astronomical cycles was strengthened by many studies. The main contribution came from a group of researchers from five universities, headed by John Imbrie. They called themselves SPECMAP, from the name 'spectral mapping' – the 'spectrum' of a pattern of variations is what reveals the presence of regular, periodic contributions to the overall picture of confused ups and downs of, in this case, temperature. The biggest surprise to emerge as they extended this kind of study further back into the past, using longer stretches of sediment core from the sea bed, was the dominant role of the hundred-thousand-year eccentricity cycle. This cycle ought to be the weakest of the three astronomical influences, and produces variations in insolation amounting to no more than 0.1 per cent of the annual total. And yet, by 1982 the analysis had shown that ice age rhythms had marched precisely in step with this long-term cycle for at least the past eight hundred thousand years. The other two cycles are also present in the 'signal' from the same sediments, and together the astronomical influences account for about eighty per cent of the variation in climate over timescales between twenty and a hundred thousand years long. Of course, other influences are at work – volcanoes erupt, ocean currents shift position and so on – but the most important set of influences determining the pattern of ice age/interglacial rhythms is indeed the Milankovitch mechanism (*Figure 3.3*).

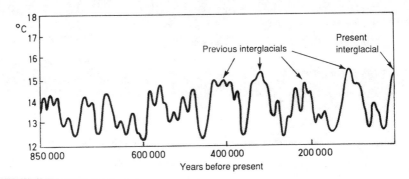

Figure 3.3
A variety of geological techniques reveals the changing level of global mean temperatures over the past 850,000 years. Interglacials, like the one in which we live, have been rare and short lived. The 'normal' state of the Earth over this time span has been a full ice age.
(Source: US National Academy of Sciences)

Some researchers suggested that the strength of the hundred-thousand-year rhythm, in particular, might be caused by a feedback involving the growth of ice sheets. Once ice sheets start to spread, they reflect away more of the incoming solar heat, and help to cool the Earth – so a small initial cooling can be amplified, provided it is big enough to encourage a small spread of ice. But these processes involve ice spreading over *land*. Even a little snow falling on land can lie around and form ice, but snow falling on the sea melts. The feedbacks *might* play a part in the development of ice ages in the northern hemisphere, where there is a lot of land at high latitudes, near the permanent polar icecap; but they cannot work in the south, where the region around Antarctica is largely ocean. Since both hemispheres march into and out of ice ages in step, there must be some other process at work.

Whatever process it was, the researchers soon realized that it provided a powerful new method of dating ancient sediments. Once they had proved the reality of the astronomical influence on climate, the changing proportions of oxygen isotopes in ancient sediments gave them a more precise tool for dating those sediments than most of the alternative techniques. And the same pattern of variations has also been found in other places, including rocks hundreds of millions of years old.

In some lakes in cold parts of the world, distinct layers of sediments are produced each spring when water from melting snow and ice rushes down from the hills, bringing a burden of sediment with it. The layers are called 'varves', and they are as distinct as the annual growth rings of a tree, but only about a millimetre thick. The exact thickness of each layer depends on the weather the year it was laid down – how much ice melted in the spring, and how soon. Over geological time, the lake bed may dry out and be squeezed and heated by geological processes to form a layer of rock, but in some cases the varve layers can still be seen today. Rocks that are two hundred million years old, from the Newark Basin in the northeastern United States, show such traces, forming patterns of variation (thicker and thinner varves) that correspond to all the astronomical cycles.

This discovery was especially exciting for astronomers, since there is no direct way of calculating the Earth's changing orbital properties back for more than a few million years – the calculations involve, strictly speaking, the gravitational influences of *all* the other planets, not just the Sun and the Moon, and although some effects can be ignored on shorter timescales, the further back you look the more important it is to include everything. Even with modern computers, it is impossible to include all of the planetary influences accurately enough to track all the orbits back in time for hundreds of millions of years. But the record in the rocks provides direct evidence that the Earth was subject to the same astronomical rhythms two hundred million years ago that affect it today – and the implication is that the whole Solar System was following the same pattern then as it does now.

But the record from two hundred million years ago also tells us that these astronomical rhythms were exerting their influence at a time when the world was warmer than it is today, and free from ice. There could not be any feedback effect from the growth of ice sheets strengthening the signal in a world where there were no ice sheets, so the puzzle of how the astronomical influences worked seemed even more mysterious.

In the 1980s, the study of astronomical cycles in the geological record went from strength to strength. Microscopic life forms in the water off the Arabian peninsula, including the ubiquitous foraminifera, are affected by temperature changes in the water, which are themselves linked with variations in the Indian monsoon. Warren Prell, one of Imbrie's colleagues at Brown University, headed a study which found that traces of the remains of these creatures in the sea floor sediments vary with a period close to the main cycle of the precession of the equinoxes – that, in other words, the intensity of the monsoon is affected by the astronomical cycles. Further studies of the sediments from this region, off Oman, showed all of the now familiar Milankovitch rhythms, and one more.

As well as the hundred-thousand-year cycle, the eccentricity of the Earth's orbit also varies with a rhythm just over four hundred thousand years long. In the first studies of the Milankovitch Model, with only a few hundred thousand years' worth of data, there had been no point in looking for this cycle, and it scarcely even got a mention. You can hardly find a repeating pattern four hundred thousand years long in a core spanning just half a million years; you need two full cycles, at least, even to suspect the pattern is there. But once the Milankovitch Model became respectable, researchers studying hundred-million-year-old rocks from central Italy found that both the hundred-thousand and four-hundred-thousand-year cycles showed up strongly in these strata, originally laid down as marine sediments. They also show up in the long-term pattern of monsoon variations. And sediments from two hundred million years ago, located near Lyme Regis in England, show the effect of the forty-one-thousand year cycle. The cycles pop up everywhere, in sediments from different regions of the globe deposited at different geological times. The Milankovitch Model, which only twenty years ago was scarcely even an also-ran among candidates to explain the pattern of past ice ages, is now supported by an impressive weight of evidence and is fully accepted as the key to understanding ice age rhythms.

But identifying the existence of the rhythms is quite a different task from understanding how they work to control climate. Even taking the three basic cycles alone, it is clear that the combined influence on the Earth is never exactly the same from one ice age (or one interglacial) to the next. Although the Earth is warm today, and it was warm about 120,000 years ago, the warmth of the two interglacials was produced by two different combinations of the three cycles. Because neither twenty-three nor forty-one thousand divides exactly into a hundred thousand years, the *exact* pattern does not repeat for literally billions of years. Fortunately, that problem need not bother us now. We are especially interested in what happened during the most recent ice age, and maybe the one before that, and on that sort of timescale there is now clear proof that the astronomical cycles exert their strong influence on the climate of our planet through an intermediary: the carbon dioxide greenhouse effect.

The carbon dioxide connection

Solid evidence that the greenhouse effect – or the lack of it – might help to explain the pattern of ice ages came in 1982, when researchers from the University of Bern, in Switzerland, reported their analysis of bubbles of carbon dioxide trapped in the icecaps. When snow falls on the ice, it is light and fluffy and there are plenty of gaps between the snowflakes. These gaps are full of air. As more snow falls and the years pass, each layer is squeezed down in its turn and becomes ice, but some of that air stays trapped in the ice as bubbles. On a polar icecap (or, indeed, on a mountain glacier) the youngest ice is at the top, and successively older layers lie deeper below the surface. Because the snow is laid down in regular annual layers, which are still visible when it has been compressed into ice, the age of a piece of ice from a particular depth can be determined, in some cases, by drilling a core of ice from the sheet, and counting the layers down from the surface. The technique, pioneered by a Dane, Willi Dansgaard, and his colleagues in Copenhagen, is useful for climatologists because the same kind of isotope measurements that reveal temperatures of long ago from the deep sea sediments also reveal how temperature has changed over the period of time covered by the ice core (more of this aspect of Dansgaard's work in Chapter Four). But the Swiss researchers decided to investigate not the temperature record in their core sample, but the atmospheric record.

Similar efforts had been made before, but with mixed success; it was the Swiss work that opened the eyes of climatologists to the implications of the technique. Careful studies of the air trapped in the bubbles in the ice showed that about twenty thousand years ago, when the most recent ice age was at its coldest, the amount of carbon dioxide in the air was only in the range from 180 to 240 parts per million, compared with 280 ppm in the early nineteenth century, before widespread burning of fossil fuel began. About sixteen thousand years ago, at the time when, we know from other geological evidence, the ice sheets began to melt, the proportion of carbon dioxide began to increase, and by the end of the ice age it had reached roughly the same concentration as the natural level today. The evidence was persuasive, but many climatologists were worried that carbon dioxide might somehow be leaking out of the old bubbles buried in the ice, or that the tricky techniques needed to extract and measure the tiny samples of old air might be giving a false reading. These doubts were allayed when Nick Shackleton and his colleagues found exactly the same pattern of carbon dioxide variations from a completely independent technique, using deep sea sediments and another kind of isotope study, this time looking at carbon instead of oxygen.

These studies provide a beautiful example of science at work, with insights that are important to our lives in the years ahead coming from a combination of biology, physics and chemistry applied to studies of the shells of long dead, microscopic sea creatures found in sediments hauled up from the bottom of the sea. It all starts with an understanding of how the oceans would absorb carbon dioxide if there were no life on Earth.

On a lifeless Earth, carbon dioxide would be taken out of the air and dissolved in the water of the seas, and as I mentioned in Chapter Two this could play a part in controlling global temperatures, although possibly not as effective a part as the processes involving life. If the world were physically the same as it is today, but lifeless, there would be about three times more carbon dioxide in the air than there actually is. Microscopic life forms near the surface of the ocean – plankton – act as a pump, extracting carbon dioxide from the air and depositing it on the sea floor. They take up carbon dioxide in the form of dissolved carbonate, and when they die their shells and skeletons, rich in carbonate, fall to the bottom of the sea. Some of this material is buried and becomes part of the long-term geological cycles that provide one of the feedbacks by which the temperature of Gaia is regulated. But some of the material dissolves back into carbonate in the deep water. This carbonate-enriched water, however, cannot rise back up to the surface. It is colder and more dense than the water above, and, like hot air, it is *hot* water that rises. So as well as taking some carbon dioxide out of circulation altogether for millions of years, the biological pump increases the concentration of carbonate in deep waters at the expense of surface waters and, ultimately, the atmosphere.

Now the carbon isotopes come into the story. Like oxygen, atoms of carbon come in different varieties. The important ones, in this study, are carbon-12 and carbon-13. Both are stable, and carbon-12 is the lighter of the two isotopes. Because of the way their biochemistry works, the plankton prefer to take up the lighter isotope to build their shells. So the biological carbon pump not only shifts carbonate to the bottom waters, it also preferentially shifts carbon-12 rather than carbon-13 out of the surface layers. Some organisms, however, live in the cold, deep waters, not near the surface. They build their shells out of the carbonate that is available to them, and this is already enriched with light carbon, so their shells contain even less carbon-13, proportionately, than the shells of creatures that live in the surface layers. Experts can identify the two kinds of shells, even in sediments tens of thousands of years old, and can sort them into the two categories. When the two types of shell are analysed, they provide a measure of how the amount of carbon being shifted from the surface to the deep ocean has varied – carbon-13, rather than carbon-12, is actually used as the indicator, because there is less of it in the first place, so small variations in the balance between the two isotopes show up more strongly as a fraction of carbon-13 than of carbon-12. As the activity of the biological pump changes, so the proportions of carbon-13 being locked away in these shells changes; by measuring the carbon isotope ratio in shells from both surface creatures and bottom dwellers with exquisite precision, the researchers can determine how active these biological processes were at different times in the past, and infer how much (or how little) carbon dioxide can have been present in the air in order to match these measurements. If there were no biological activity, for example, then the ratio of isotopes would be the same everywhere, in the surface and the deep sea, and we would know that the carbon dioxide concentration was about three times the present value.

By now, it should come as no surprise to find that these studies show that there was less carbon dioxide in the air during the latest ice age than there is today. The ice core studied by the Swiss team extends back over the past forty thousand years, and over that timespan the concentrations of carbon dioxide measured in the ice core bubbles and the concentrations inferred from isotope measurements of sea bed sediments are exactly the same. But the sedimentary record goes back much further, for hundreds of thousands of years, and provides much more information. The intriguing new discovery is that the changes in carbon dioxide *precede* the changes in ice cover. Working with Nicklas Pisias, of Oregon State University, Shackleton applied the technique to analyse a sediment core covering a span of 340,000 years, more than three complete glacial/interglacial cycles. They found all three of the Milankovitch rhythms present in both the temperature record (determined from oxygen isotopes) and the carbon dioxide record (determined from carbon isotopes) in the core. But by comparing their findings with calculations of how the astronomical cycles have varied, they found that the changes in the Earth's orbit precede the changes in carbon dioxide, and that the changes in carbon dioxide in turn precede changes in climate – or, at least, changes in ice cover. Somehow, the astronomical changes cause the biological pump to increase or decrease its activity, and this produces a change in the strength of the greenhouse effect which is then a key element in switching the world into or out of an ice age – and which also helps to explain ups and downs of temperature on lesser scales, *within* an ice age or an interglacial. Carbon dioxide amplifies the changes that the Milankovitch process is trying to produce.

The effect almost certainly operates through changes in the workings of the carbon pump at high latitudes. High latitudes feel the effects of the Milankovitch rhythms more strongly than tropical regions do, and it is also at high latitudes that the deep ocean water breaks through to the surface and makes direct contact with the atmosphere. Today, there is not enough sunlight reaching high latitudes, some researchers suggest, for the plankton that live there to use all of the available nutrients; if there were more sunlight, as there is at some other times in the Milankovitch cycles, they might grow more vigorously, taking carbon out of the atmosphere as they do so. All such proposals, however, are still very tentative. The only way to get a better understanding of the links between astronomy, biology and climate that control the pulsebeat of ice ages is by taking a more detailed look at the latest ice age. Almost as if it had been waiting in the wings for its cue, just such a more detailed picture of recent climatic changes emerged in 1987 from a painstaking analysis of ice samples (and bubbles trapped in them) from a core more than two kilometres long, drilled through the Antarctic ice cap.

The Antarctic cornucopia

The core was drilled at the Soviet Union's Vostok Station, in East Antarctica. Drilling began in 1980, and eventually reached a depth of 2.2 kilometres. The ice at the bottom of the borehole is formed from snow that fell in Antarctica more than 160,000 years ago, in the ice age before last. At that depth, a year's snowfall from that long gone era corresponds to a layer of ice just one centimetre thick. It is hard for anyone who has never been involved in such a project to comprehend the engineering skills required to extract a core of ice this long, especially bearing in mind that the job was being carried out in the most inhospitable continent on Earth. The best measure of the difficulty of the task is that it took from 1980 to 1985 to extract the core. That borehole has now been abandoned, squeezed shut by the inexorable pressure of the ice, but there are plans for a new hole, reaching three kilometres or more down into the ice.

The Vostok core has been analysed by a large team of researchers in a collaboration involving scientists from France and the Soviet Union. Their main conclusions, based on analyses of the first 2,083 metres of core extracted from the hole, were reported in the journal *Nature* in October 1987; but new insights will continue to emerge in the years ahead, as the pieces of the core, carefully preserved in cold storage, are subjected to new tests.

Today, the average surface temperature at the Vostok site is −55.5°C The Franco-Soviet team used yet another isotope technique to infer past temperatures from their analysis of the core. This time, the 'thermometer' was based on the concentration of heavy hydrogen (also known as deuterium) in the ice. The principle, however, is the same as with the oxygen isotopes – heavier molecules of water (ones that contain deuterium) evaporate less easily but freeze out of water vapour in the air more easily, so the amount of heavy hydrogen in snow falling in Antarctica to be preserved as ice depends on temperature. Deuterium provides a direct measure of the temperature when and where the snow was falling, while the oxygen isotopes give a broader picture of global changes in ice cover. The deuterium record, in fact, closely matches the oxygen-18 record from ocean sediments, but the ice cores provide more detail. The analysis shows that at the coldest points of both the latest ice age and the penultimate ice age temperatures were 9°C lower than they are today, while during the warmest years of the interglacial that separated the two latest ice ages, a little over 150,000 years ago, temperatures were 2°C higher than they are today.

Samples of air from bubbles trapped in the ice show that there was a dramatic increase in the concentration of carbon dioxide in the atmosphere, from 190 ppm to 280 ppm, at the end of each of the two latest ice ages, and there was a comparable decrease in the carbon dioxide concentration at the beginning of the most recent ice age (*Figure 3.4*). The temperature record, in particular, also shows clear signs of the familiar Milankovitch cycles at work.* But as well as confirming the existence of connections between carbon dioxide, ice ages and the Milankovitch rhythms, ice cores from both the Antarctic and the Arctic are providing completely new insights into the climatic changes that cause the shift from ice age to interglacial and back again; those new insights point to a key contribution from the Earth's biological systems – a role for Gaia.

The Gaian connection

The evidence comes from a surprisingly small number of cores. Over recent decades, scientists have analysed more than twenty thousand cores from the deep sea, and many short ice cores. But they have drilled and studied in detail only five deep ice cores, two from Greenland and three from Antarctica. The Vostok core

---

*One complication thrown up by these studies is that they suggest that in Antarctica, at least, at the *beginning* of an ice age (and, remember, even the Vostok core is only long enough to show the beginning of *one* ice age), the temperature changes just before the carbon dioxide concentration changes. This is different from the deep sea sediment analysis, which spans several ice ages and interglacials, and shows that *global* temperatures change after the carbon dioxide concentration changes. When more ice cores are analysed and these geographical details are better understood, they may be able to tell us more about how the Milankovitch mechanism works.

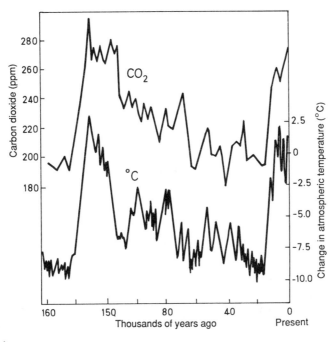

Figure 3.4
As the Earth shifts into or out of an interglacial, temperature variations and changes in the concentration of carbon dioxide in the atmosphere move in step. (Data from the Vostok core)

is the deepest, but all five provide invaluable information about climates of the past. This information is so valuable that participants at a Dahlem workshop on the environmental record in glaciers, held in Berlin early in 1988, described the ice cores as almost 'an ideal recorder' of the Earth's part environmental and atmospheric conditions. They contain traces of both particles and gases from long ago, independently record temperature changes, store the information in datable layers, and preserve the record free from contamination. This record is increasingly seen as indicating a biological connection with climatic change.

In 1988, for example, the Vostok team reported that the amount of methane in the atmosphere almost doubled at the end of the penultimate ice age, between about 150,000 and 130,000 years ago. Methane is another important greenhouse gas, and other ice core studies have shown that its concentration in the air has increased from a natural background level of about 0.7 ppm three hundred years ago to about 1.7 ppm today, presumably as a result of human activities (more of this in Chapter Six). The Vostok study showed that during the previous interglacial the concentration of methane in the atmosphere was about 0.62 ppm, fairly close to the natural concentration that seems to have been present during the present interglacial before human intervention. But even longer ago, during the penultimate ice age, the concentration was only 0.34 ppm. The increase in methane at the end of that ice age would have provided one quarter as much greenhouse effect warming as the increase in carbon dioxide that occurred at the same time, and the build-up of the two greenhouse gases together can account for about half of the warming that occurred as the ice melted: 2.2°C out of 4.5°C.

Nobody can be sure why the methane concentration rose at that time, but we can make an educated guess. The bacteria that produce methane live in swampy wetlands, and it seems likely that either the bacteria became more active as the temperature rose, or that as the world warmed and ice melted there was more swamp around for them to live in. Whatever the details, though, since methane is almost entirely produced by biological activity, this is a clear example of a biological feedback acting to enhance a climatic shift. Those bacteria, over a period of about twenty thousand years, strengthened the greenhouse effect by half the amount that human activities have strengthened it over the past hundred years.

That helps to explain why the world warmed so much at the end of an ice age. Gaian influences can also explain the sharpness of the cooling that occurs when an ice age begins.

Several ice core studies have investigated how the amount of dust in the atmosphere changes when the world's climate changes. Fine particles of dust, known as aerosol, can drift around the world at high altitudes in the atmosphere acting as a sunshield, veiling the Sun and cooling the Earth below. Such dust veils are produced by volcanic activity (which also produces a haze of sulphuric acid droplets in the stratosphere), and may account for short-term dips in temperature following major volcanic eruptions. They are also produced from particles blown in the wind from dry regions of the globe. Whatever their origin, the particles eventually fall out of the atmosphere, and some of them fall with the snow at high latitudes. Aerosol particles are found in different numbers at different depths in the ice cores, showing that the strength of the dust veil effect has varied over the millennia. In particular, the ice cores show that there was more atmospheric aerosol at the coldest time of the most recent ice age, the 'Last Glacial Maximum', or

LGM, that occurred about eighteen thousand years ago.

This dust must have played a direct part in cooling the world, through the veiling effect. When the world is cooler, there is less evaporation from the oceans, so there is less rainfall and the land dries out. So once a cooling sets in it is easy to see how dry winds may blow more dust off the land to build up the aerosol layer – another positive feedback process. The ice core studies show that whereas carbon dioxide changes occur first, leading the world into or out of an ice age (in response, of course, to the astronomical cycles), the dust increases after the cooling has already begun, amplifying the process. But even the dust plays a part in biological activity.

One of the strangest discoveries from the ice core analyses of the late 1980s was the way in which traces of methanesulphonic acid vary from layer to layer in the Antarctic ice. Methanesulphonic acid, or MSA, is produced from dimethylsulphide, which is itself one of the waste products of the micro-organisms that inhabit the surface layers of the ocean (it is dimethylsulphide, in fact, that gives sea air its distinctive smell). During the most recent ice age, the amount of MSA being deposited in Antarctica each year was between two and five times greater than today, indicating that those organisms were more active then. This was a timely discovery for proponents of the Gaia hypothesis, because Jim Lovelock and his colleague Bob Charlson, of the University of Washington, Seattle, had already suggested that the dimethylsulphide (DMS) produced by biological activity in the surface layers of the ocean could modulate climate by causing changes in cloud cover.

In its original form, the idea was proposed as a way in which Gaia could prevent the Earth from becoming too hot or too cold. DMS produced by microscopic life forms in the ocean gets into the atmosphere, where it reacts to produce particles that can act as the 'seeds' on which the water droplets or ice particles that make clouds can grow – the seeds are mainly, in this case, sulphuric acid droplets. If there were fewer clouds around than there are today, said Lovelock and Charlson, more sunlight would reach the surface of the sea. This would make the world hotter, but it would also stimulate biological activity, since even microscopic plants need sunlight for photosynthesis, releasing more DMS and causing cloud cover to spread, reflecting away sunlight and cooling the globe. If the cloud cover spread 'too much' for Gaia's comfort, the lack of sunlight would make biological activity drop, DMS production would fall, and there would be fewer clouds so the world would warm. The system is an example of a negative feedback, always acting to restore the status quo when it is disturbed.

One of the key features of the idea is that it will only work if the atmosphere of the Earth, at least over the oceans, is in some sense 'ready' to form clouds today, but lacks the seeds on which they can grow. The fact that the potential for more cloudiness is indeed there has been confirmed by satellite observations, which show that clouds grow along the tracks taken by ships traversing the Pacific Ocean. These clouds are seeded by pollution from the funnels of the ships, and measurements of the reflectivity of the clouds show that they do reflect away heat that would otherwise reach the surface. What pollution from ships can do, said Lovelock and Charlson in effect, DMS can do at least as well. But they had no proof; the link between cloud cover and Gaia seemed like a flight of fancy that might never be pinned down as having practical relevance, until those measurements of biological sulphate from the Antarctic ice were reported. Charlson and his colleagues quickly calculated the size of the change in cloud cover implied by the change in biological activity determined from the sulphate variations, and found that the increase in cloudiness during ice age conditions could have produced a cooling of the globe by 1°C – not quite as big as the cooling due to the reduced carbon dioxide greenhouse effect, but much bigger than the change produced by the difference in methane concentration of the atmosphere.

Clearly, the carbon dioxide and DMS effects should march in step. Increasing biological activity among the plankton would take carbon dioxide out of circulation, and at the same time increase the amount of DMS available to seed clouds. The organic sulphate only gets into the Antarctic ice, after all, because it has once been in the clouds from which snow has fallen. But why should the plankton thrive when the world begins to cool?

John Martin and Steve Fitzwater, of the Moss Landing Marine Laboratories in California, may have at least part of the answer. They found it during a study which jumped off from a long-standing biological puzzle. The waters of the Antarctic Ocean and the sub-Arctic Pacific Ocean are rich in important plant nutrients, such as phosphates and nitrates, that are essential for plant growth but which are not being taken up by the microscopic plants that live in those oceans. Clearly, some other essential ingredient required by the plankton is missing from those waters, and limiting their growth. But what could the missing ingredient be?

The two researchers found out by collecting samples of ocean water from the northeast Pacific and adding iron, in the form of dissolved iron compounds, to them (Martin had already guessed that iron might be the key). When dissolved iron was added to the samples, rapid plankton growth began after four days, and there was a substantial increase in the amount of chlorophyll in the water, while all of the available nitrogen was used up. Iron is an essential ingredient for many life processes, and in particular it is a component of chlorophyll so there is no difficulty understanding these results. The studies show that what is stopping the plankton from increasing in the cold, nutrient-rich waters at high latitudes is a shortage of iron.

Which brings us back to wind-blown dust. During ice ages, there is more dust in the air, as dry winds blow over land masses which have less rainfall than today. Once such a process begins – for whatever reason – the dust-laden air will carry compounds containing iron over the oceans, providing the plankton with the essential ingredient they need to take advantage of the available nutrients, stimulating photosynthesis and not only taking carbon dioxide out of the air but also producing DMS to seed cloud cover.

The start of the cooling, triggered by the Milankovitch mechanism, may well involve an increase in the amount of sunlight available at high latitudes, which encourages plankton to grow up to the limit of their ability without added iron. This would then trigger a positive feedback, a runaway in which cooling released more iron from the land and caused more cooling, until the growth of plankton stopped because they ran out of other nutrients. The process might go into reverse when conditions changed so that there was too little sunlight available for the plankton to continue to be so active. A reduction in their activity would leave more carbon dioxide in the air, helping to warm the world, and thereby increase evaporation from the seas, and damp down the dusty continents. Less wind-blown dust would mean less iron for the plankton, and less activity among the high latitude phytoplankton. The broad outlines of how such a feedback might work are clear; but because the changes in sunlight caused by the astronomical cycles have the opposite effect in opposite hemispheres, a great deal more work needs to be done.

These are new ideas, and the details are bound to be amended and improved over the next few years. But they are important to the tale of how human activities are now affecting the climate because they show the complex ways in which living and non-living components of the environment interact. Although we can calculate fairly accurately how much the world will warm if we simply double the amount of carbon dioxide in the air, it is impossible to do more than make educated guesses about how and when biological systems will be affected by such a change, and how (and when) they in turn will affect the climate, perhaps strengthening the greenhouse effect or, perhaps, acting to reduce the speed with which the Earth warms. But while it might be comforting to hope that Gaia might resist a sudden warming, the evidence in the ice is less reassuring. About eleven thousand years ago, at the end of the most recent ice age, the temperature jumped by five or six degrees Celsius in the span of forty years – and there were similar short-term fluctuations during the ice age itself. One of the most disturbing implications of all this work, however, is the clear indication that by adding carbon dioxide and other greenhouse gases to the atmosphere today, human activities are going against the natural trend. Left to its own devices, the climate cycles show, the world would now be beginning to settle down for the next ice age. For the past three million years, at least, the natural state of the world has been in an ice age. Gaia only emerges from the ice age briefly, almost reluctantly, when all three Milankovitch rhythms conspire to make the world warmer.

From a human point of view, especially if you live in temperate or higher latitudes, it is easy to imagine that a warmer world would be a much better proposition than a new ice age. But all the evidence shows that there was a lot more biological activity, in the oceans at least, when the world was cooler. Plankton may not mean much to you and me – they are hardly romantic, cuddly creatures like pandas (or, rather, like the public image of the panda) and they are not the tragically doomed, intelligent creatures that we imagine whales to be. But most of 'our' planet is covered by water, and most of the Earth's biomass lives in the sea. Judging from the record of past biological activity in the ice cores, that biomass does better when the world is colder. Gaia may actually *prefer* an ice age to the present climatic conditions; but we can see how far, and how rapidly, we are now shifting the world out of its natural pattern by taking a look at the relatively modest natural fluctuations in climate that have occurred since the latest ice age, and at the relatively large influences of those climatic shifts on our immediate ancestors.

FOUR

After the Ice

The closer we come up to date, the more detail we can read from the geological record of past climates. At the end of the latest ice age, some ten thousand years ago, dates are still a little vague and exact temperatures cannot be inferred from the records. For the past three hundred years, dates are known precisely, and temperatures and other meteorological parameters are recorded, at least for some parts of the world, by competent scientists who made the measurements on the spot. One thing that is clear, though, is that the warmest centuries of the entire interglacial to date occurred soon after the ice melted. This might seem surprising, if you expect things to warm up the further we get from an ice age; but from another perspective it makes perfect sense, since it was only the fact that the world warmed significantly that caused the ice to retreat. At one time, climatologists used to refer to the peak warmth, around six thousand years ago, as the 'climatic optimum', on the assumption that a warmer world would be a more pleasant place to live. Partly in view of present concern about the hazards of a greenhouse effect warming of the globe, that term has fallen from favour and the more common expression used now, based on the geological name for the post-glacial epoch, is the 'Holocene warm period'. It is also known as the Altithermal, or hypsithermal. The shift from ice age into interglacial is recorded in the rocks and sediments of the Earth in the form of debris left by retreating glaciers, changes in the geographical distribution of trees and other plants (and, therefore, of the seeds left behind for us to study now), changes in sea level (shown, for example, by traces of old beaches) and other indicators. Isotope studies can give only a broad picture of global changes in sea level and temperature, but these more traditional tools of 'paleoclimatology' can tell us about regional changes. When the information from various sources is pulled together, it shows that the ice did not retreat everywhere at the same time, and that the shift into interglacial conditions actually took several thousand years.*

Warm and wet/cold and dry

The last glacial maximum, a peak (if that is the right word) of intense cold, occurred roughly eighteen thousand years ago. For the convenience of having a round number to refer to, the end of the latest ice age, and the beginning of the present interglacial, is usually set at ten thousand years ago, but even parts of northern Europe were still more icy then than they are today, while in North America the retreat of the ice had barely begun in earnest. The ice sheet over Scandinavia finally disappeared about eight thousand years ago, when the North American ice sheet was still half as big as it was during the last glacial maximum. Ice persisted over Labrador until about 4,500 years ago, right through the Holocene warm period, which really only deserves its name (in the northern hemisphere) if you look at the records from Europe, Asia and Africa. Eastern North America was kept cool by the proximity of ice, and cold air blowing down out of the north kept summers chilly there until 4,500 years ago.

This difference between the North American continent and the rest of the world is both important and unfortunate. Important, because it had an influence on the entire circulation of wind and weather patterns around the northern hemisphere; unfortunate, because that means that we cannot hope to understand what the warmth of the next few decades will be like simply by studying the Holocene warm period, since there are no permanent ice sheets covering Labrador today. As the experts succinctly put it, the boundary conditions were different then. Even so, the Holocene warm period does give some idea of how much dif- ference a warming of a couple of degrees Celsius means to life on Earth.

At the peak of the warm period, roughly six thousand years ago, the forests in the west of both Canada and Siberia pushed

*A definitive 'pulling together' of this kind is in Hubert Lamb's epic book *Climate: Present, Past and Future* – see the Bibliography for further details.

70

two or three hundred kilometres further north than they do today. Summer temperatures in Europe and North America must have been between two and three degrees Celsius warmer than they are now, and the waters between Japan and Taiwan were about 6°C warmer than they are at present. Although a colder period set in on the continents, especially in Canada, a little less than five thousand years ago, the Arctic Ocean was at its warmest around 4,500 years ago when the *northern* fjords of Greenland were free from ice, and driftwood from Siberia reached those shores, above 83°N.

Between about seven and five thousand years ago, sea level rose rapidly around the world as the last remnants of the great ice sheets melted, and by about four thousand years ago it was about three metres higher than today. That extra height corresponds to the volume of water produced by melting a million cubic kilometres of ice – there was that much less ice cover over the land then than there is today, and there has never been less permanent ice cover at any time during the present interglacial. The contrast with the ice age that had just ended was sharp, and rapid. Nine thousand years ago, there was still so much water locked up in ice that a vast, dry plain linked England to Europe across the bottom of what is now the North Sea, and a similar link joined Wales and Ireland. It would have been possible to walk from the west of Ireland all the way to France (and, indeed, on to China). About that time, perhaps a little earlier, the Black Sea was still an inland lake, cut off from the Mediterranean. But by eight thousand years ago, Britain and Ireland were both islands (the water arriving in time to stop the spread of snakes up from the south into Ireland, but not before they had crossed into England) and the Black Sea was, indeed, a salty sea.

One big difference between the climate of the Holocene warm period and the pattern we can expect in the twenty-first century is that the Sahara and some of the present day deserts of the Middle East were wet. The tropical rain belt pushed northward as the world began to warm, but the moist winds further north, that bring winter rain to the Mediterranean region and to California today, still extended their influence southwards as long as the traces of ice sheets remained. The desert was squeezed from both sides. Africa and the Middle East experienced more widespread monsoon rains in summer, but ample winter rainfall as well. The influence extended to Mexico, and to India, where the edges of what is now the Tharr desert, with rainfall of only 250 millimetres a year, had

71

at least twice as much rain as today – and perhaps three times as much – each year. In the southern hemisphere, eucalyptus flourished in parts of southwest Australia that are now dry. The world was both warmer and wetter six thousand years ago.

Then, the pattern changed – why, nobody knows. Temperatures turned down after about five thousand years ago, and the regions that are deserts today, from Australia to India, Mexico and the Sahara, began to dry out. Glaciers advanced once again on the mountains in Europe and Alaska, and probably over the Rockies, and sea level fell back from the heights it had reached in the Holocene warm period. By about 2,500 years ago, temperatures in Europe were about 1°C lower than they had been in the warmest post-glacial centuries (but still warmer than they were in the middle of the twentieth century). Like the shift out of the ice age, however, there were geographical differences. Northern Europe became wetter, and peat bogs spread across Ireland, Scandinavia and Germany, while moisture-loving trees spread across Russia and Canada. All these changes had a pronounced effect on the people who were alive at the time, and on civilizations that had begun to flourish during the Holocene warm period. Taking just one example,* Hubert Lamb points out that glaciers advanced in Norway, until about 2,700 years ago they extended almost as far as during the coldest centuries of the past millennium. He says that it is generally accepted by the experts that this period of increasing cold and ice provided the foundation for the stories that became the legend of Ragnarok, the twilight of the gods, which represents a folk memory of a former, pleasant way of life in the north being destroyed by the spread of ice – the dreadful Fimbulvinter, described in the epic poem 'Edda' as the time when 'the snow drives from all quarters with a biting wind; three such winters follow one another and there is no summer in between'. Against that background, 'optimum' doesn't seem such a bad description of the Holocene warm period, after all! Just as that warm period was succeeded by a cold spell, the cold epoch in turn gave way to another warm interval, and that warming was followed by another cooler interval. But the secondary warming of the present interglacial never reached the warmth of the Holocene warm period, while the cold spell that followed was so severe that it has become known as the 'little ice age'. By now, we are talking

*If you would like more, check out *Children of the Ice*, by John and Mary Gribbin (Blackwell).

about changes that have occurred within the past two thousand years, during the Christian era; and we have another natural thermometer with which those changes can be assessed; the isotopic equivalent of a fine-toothed comb.

A little ice age

The details come from an ice core which was, in fact, one of the first to be analysed by the oxygen isotope technique. It was drilled from a site known as Crête station, in the heart of the Greenland glacier, and analysed in detail by Willi Dansgaard and his colleagues in the early 1970s. Preserved in cold storage, the ice samples are still available to provide new information today, as new techniques are developed to unlock their secrets; but the first, and still the most important, piece of information to be extracted from the core was a temperature curve covering a span of 1,420 years, back in time from the 1960s. The particular beauty of this is that because the 404-metre-long core comes right up to date, and the layers of ice in it can be identified year by year, the calibration of isotope ratios against temperatures can be set precisely for recent decades, when temperatures were being measured by other means and recorded in scientific journals. So the temperatures inferred from isotope ratios in samples of the ice from centuries past provide a record that is both accurate and continuous over this long span of historical time.

The two main climatic features of this time span that I have already mentioned stand out clearly in the Greenland isotope record (*Figure 4.1*).

From about 900 AD to 1100 AD, the world was warmer than it had been since the Holocene warm period, and warmer than it would be again until the end of the twentieth century. It wasn't quite as warm as the Holocene warm period itself, so it is sometimes known as the 'little' climatic optimum, although the term 'Medieval warm period' is usually preferred by climatologists. Many pieces of evidence, not just the Greenland ice core, show that this warm interval was like the Holocene warm period, but less prolonged and less pronounced. Icebergs and pack ice were restricted to higher latitudes than during the cold period (a major reason why the Norse seafarers were able to explore the North Atlantic, colonizing both Iceland and Greenland, and reaching North America). In western and central Europe, vineyards extended three to five degrees further north, and up to two hundred metres higher above sea level

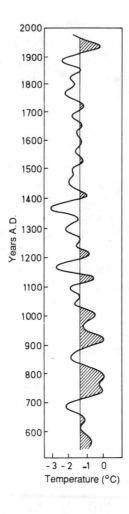

Figure 4.1
Changes in the proportion of oxygen-18 in samples from the Crête core drilled in
central Greenland can be interpreted as a record of temperature variations going
back to the sixth century.
(Adapted from a figure supplied by Willi Dansgaard and used in Climatic
Change *(edited by John Gribbin; Cambridge University Press, 1978))*

than today – evidence that temperatures were about 1°C higher
than those of the middle of the twentieth century.

But the little optimum soon came to an end. It was followed
by the cold of the little ice age, centuries of severe weather that
set in before 1200 AD and persisted, with some variations, until
1900. Two other ice cores from Greenland, one going back to 1232
AD and one to 1176 AD, confirm the general run of temperatures

revealed by the longer Crête core, and for the later part of this period there is ample historical evidence about the severity of the little ice age in Europe, and some from other parts of the world, notably China. The historical records do not precisely match the temperature changes indicated by the isotope variations in the Greenland cores. In particular, although the coldest century of the little ice age seems, from the Crête isotope data, to have been between about 1300 AD and 1400 AD (*Figure 4.1*), all the historical evidence shows that the little ice age was at its most bitterly intense in Britain and western Europe about 250 years later, in the seventeenth century. Between 1550 and 1700, Europe experienced its coldest run of weather since the ice age proper, and Lamb points out that this is the *only* period of time since the latest ice age for which evidence from *all* parts of the world points to a colder climate than that of the middle of the twentieth century. By the 1760s, temperatures were being measured (in a few places!) using thermometers, and we no longer have to rely entirely on 'proxy' records, such as ice core isotopes. In the late seventeenth century, annual mean temperatures in England were 0.9°C lower than the average for 1920–60, and in the decade of the 1690s the equivalent deficit was 1.5°C.

Different researchers attach the name 'little ice age' to different centuries. The six hundred years from 1300 to 1900 were colder than the six hundred years from 700 to 1300, and represent the maximum extent of the little ice age. Some people use the name only for the severe cold in Europe from 1550 to 1700. Others set the start of the little ice age a little earlier, and the end as late as 1850 or 1900. Wherever you set the marker, though, there is no doubt of the impact of these changes on people. The extreme weather conditions of the fourteenth century caused famines in which many people starved and whole populations were so weakened that they fell easy victims to the Black Death which ravaged Europe shortly after; drought in the 1660s helped to create the conditions which spread the last great plague to affect London, and then contributed to the destruction of a large part of the city in the great fire of 1666. There were famines in west Africa in the late seventeenth century, with summer rainfall limited to the region close to the equator; and there were frequent failures of the monsoon in India

Even within the little ice age, though, there was variation from year to year and season to season. Not every year produced summer drought and winter blizzards in England; sometimes, the monsoon arrived on cue in India. Against this background of natural variation

75

in climate and weather, it is easy to guess that the little ice age itself, and the little optimum that preceded it, were also just caused by random fluctuations in the climate system, bringing a run of warm centuries followed by a run of cold centuries merely by chance. Intriguingly, though, when the isotope record from the Greenland glacier is subjected to the same kind of statistical investigation that shows the presence of the Milankovitch cycles in longer isotope records from both the sea bed cores and the Antarctic ice, it, too, shows a pattern of regular rhythms interacting with each other and with some genuinely random variations from year to year, to produce the complex pattern of temperature variations shown in *Figure 4.2*. And just as the ice age rhythms can be related to changes in the Earth's orientation as it orbits the Sun, so these smaller, more rapid fluctuations in climate seem to follow changes in the activity of the Sun itself.

Cycles of ice and fire

One of the shorter ice cores investigated by Dansgaard's team provides the clearest evidence of this pattern at work. This core comes from a site further to the west than the Crête core, from a location known as Camp Century. Here, the isotope record matches more closely the detailed historical pattern of warm and cold centuries in Europe, with the coldest interval indicated by the isotopes being between about 1600 and 1700. Whatever the patterns of wind and weather that make for these minor geographical variations, the fact that Camp Century seems to experience temperature changes in step with Europe (and also, analysis of recent decades shows, with North America) makes it more useful for this kind of study, even though it does not extend back as far in time as the Crête core.

The two regular variations that are revealed by statistical analysis of the Camp Century isotope data are close to eighty years and 180 years long. *Figure 4.2* shows the regular rhythm produced by these two variations beating together, alongside the actual Camp Century isotope measurements, which, of course, include the effects of random variability of weather as well. When the results of this analysis were first announced, in the early 1970s, they caused a minor flurry of interest among the experts. But perhaps the interest should have been greater. First, because simply by extending the pattern of the Camp Century record into the future we can see that it implies

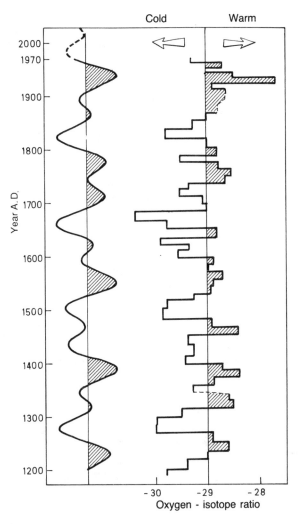

Figure 4.2
A core from Camp Century gives a shorter record than the Crête core, but this one
(right) *closely matches the known temperature variations in Europe. When these*
variations are smoothed out and analysed (left), *they seem to fit a pattern built*
up from two smoothly varying temperature cycles, one 180 years long and one
eighty years long.
(Source: Dansgaard)

a run of cold decades as we move into the twenty-first century;
secondly because these two rhythms, eighty and 180 years long,
are already familiar to astronomers who study variations in the
Sun's activity.

The fact that this evidence for a double pulsebeat of minor
climatic variations should be taken seriously is highlighted by

studies of tree rings in wood from Japan. Trees take in water from their surroundings as they grow, as well as breathing carbon dioxide, and they store oxygen atoms in the cellulose of their wood. Like oxygen atoms locked up in the ice over Greenland or Antarctica, those atoms contain different isotopes in a ratio set by the temperature during the year that the tree was adding a particular growth ring to its girth. Leona Libby and colleagues at the University of California, Los Angeles, established that the isotope ratios in tree rings can indeed be used as another proxy thermometer. This is not completely independent of the ice core thermometer, since it uses essentially the same proxy – the ratio of oxygen isotopes – but it has the advantage that it can be used at temperate latitudes, where trees grow, not just in the polar regions. Libby's studies of samples of wood from long-lived trees from Japan show just the same eighty and 180 year cycles that show up in the Camp Century record. Which is where the Sun comes into the story.

Our Sun is not a perfect, constant source of heat and light, as the ancients believed. As recently as the seventeenth century, Galileo got into trouble with the Church authorities for, among other things, pointing out that the surface of the Sun is sometimes marked by blemishes, known as sunspots; in fact, these dark spots were known to Chinese and Greek astronomers more than two thousand years ago. Observations since Galileo's time have revealed that the number of spots on the surface of the Sun varies over a cycle roughly eleven years long. Today, even at the minimum of the Sun's cycle of activity there might be a few sunspots visible. The numbers then increase from year to year until they reach a maximum (the next solar maximum is due in 1990) and then fall away again to another minimum. The interval from one minimum to the next is eleven years on average, but an individual sunspot cycle, as it is called, may be a couple of years shorter, or a couple of years longer, than the average.

The strength of each sunspot cycle – the number of spots visible at solar maximum – also varies from one cycle to another. Ironically, for a hundred years after Galileo drew attention to sunspots early in the seventeenth century there were very few such spots seen on the surface of the Sun even at solar maximum; for all practical purposes, there was no solar maximum at all from 1650 to 1710. In the eighteenth century, however, the eleven-year rhythm of solar activity became firmly established, and has lasted up until the present day.

Sunspots are now recognized as simply the most obvious and

visible features of a broad change in solar activity over the eleven-year cycle. This involves changes in magnetic activity, in the number of particles shot out from the Sun across space (solar cosmic rays) and possibly, according to the latest observations from space, very small changes in the amount of heat radiated by the Sun. The 'coincidence' that the Sun's cycle of activity seemed to have switched off exactly during the coldest decades of the little ice age had already attracted attention even before the eighty and 180-year climate rhythms were revealed by the Greenland cores; the coincidence looked even more striking in the late 1970s, when Jack Eddy, of the US National Center for Atmospheric Research, carried out a major study of similar variations going back into the Bronze Age.

Of course, no Bronze Age astronomers were around writing down observations of sunspot numbers for Eddy to analyse thousands of years later. Once again, the research depends upon proxy records. This time, the record of solar activity is provided by an isotope of carbon, carbon-14 that is found in the wood of living trees, and also in wood from long-dead trees. Unlike the isotopes I have discussed so far, carbon-14 atoms are *not* stable. They are radioactive, and are, in fact, only present on Earth today because they are manufactured in the atmosphere by an interaction between cosmic rays and nitrogen atoms. Because carbon-14 atoms are unstable, they decay into other atoms with a characteristic timescale. Some, but not all, of the atoms of carbon-14 in a piece of wood decay in this way each year, in a regular and predictable manner. The same proportion of radioactive atoms disappear each year. So the amount of carbon-14 in a sliver of wood from a tree ring of a certain age depends on two things: on the activity of the Sun in the year that tree ring was being laid down, and on the age of the sliver of wood.* If you know the age of the wood, it is easy to calculate how active the Sun was that year.

*Because the carbon-14 atoms are actually made from interactions involving cosmic rays from deep space, the relationship with solar activity is the opposite of what you might expect. When the Sun is *more* active, the wind of particles it sends streaming out into space shields us from incoming cosmic rays, and so less carbon-14 is produced in the atmosphere. When the Sun's activity is weakest, more cosmic rays from deep space reach the Earth, and more carbon-14 is produced.

Incidentally, if you are worried that all this activity might upset the isotope thermometer, don't be. Neither oxygen isotopes nor deuterium are produced in the atmosphere by cosmic ray effects, except in tiny and totally insignificant quantities.

Tree rings, of course, can be dated simply by counting the layers of wood in a sample. There are many trees around that are three or four hundred years old, and from comparing the latest rings in the samples with astronomical observations of solar activity it is clear that the amount of carbon-14 in the wood really does provide a guide to what the Sun is doing. The technique shows that the second half of the seventeenth century was a period of increased carbon-14 production, confirming that the Sun was quiet then. But this is less than one-tenth of the story Eddy has to tell. The carbon-14 record can actually be extended back beyond 3000BC, more than five thousand years into the past and more than halfway back to the beginning of the present interglacial.

Amazingly, a complete tree-ring 'calendar' extending back this far has been built up from the wood of the bristlecone pines that live in the southwestern part of the United States. These trees live to a great age, and they do so in an isolated part of the world, high in the White Mountains, where they have not been interfered with by man. Dead wood has been lying on the dry mountainsides for thousands of years, and pieces can be dated by matching up their patterns of thick and thin rings with those of other trees. If a tree now living has been growing for, say, five hundred years, and a dead log from nearby was alive between eight hundred and three hundred years ago, then the tree-ring 'fingerprint' of the outer two hundred years' worth of growth on the log will match the pattern of tree rings in the inner part of a slim core of wood drilled from the living tree. Still older wood can be dated by comparing its outer rings with the inner rings of the log, and so on back into the past. It takes a lot of painstaking work to complete the calendar, but it has been done. The longest calendar established from living trees alone pushes back eleven hundred years into the past; one built up from overlapping samples of living and dead wood goes back beyond 3000BC.

The tree rings themselves provide an indication of how the climate has changed. That is why they form a characteristic pattern of thick and thin rings, as distinctive as a fingerprint in identifying a particular sequence of years. The pines put on fatter growth rings in years when the weather suits them. This also tells us how the warmth of the growing season (April to October) has varied in the White Mountains over the past five thousand years. There was a short, cool spell about five thousand years ago, followed by a period slightly warmer than average which lasted for 2,200 years and began in 3500BC. From about 1300BC to 200BC temperatures were generally below the long-term average, then they were slightly warmer until about 1000AD. A sharp, warm spell centred on 1200AD matches up to the little climatic optimum, and the rings also clearly show the influence of cold between 1400AD and 1900AD, at the time of the little ice age which affected Europe, China and the oxygen isotopes in the water falling as snow over Greenland.

And, of course, samples of this kind contain a record of carbon-14 variations, which Eddy has used as an indicator of solar variations. When he compared the pattern of carbon-14 variations with geological evidence for the way in which glaciers around the world have advanced and retreated over the past four thousand years, Eddy found that the two matched, as he put it, 'like the fit of a key in a lock' (*Figure 4.3*). It may be that the amount of heat put out by the Sun varies by the small amount – only about one per cent – needed to explain these climatic changes; or it may be that a more complex interaction, perhaps involving the way cosmic ray particles affect the production of clouds high

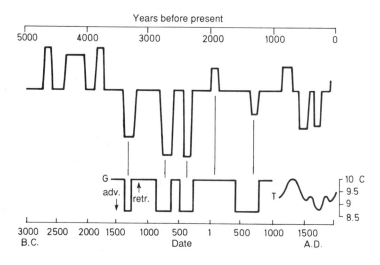

Figure 4.3
Changes in the amount of carbon-14 being produced in the atmosphere closely match the dates of advance and retreat of glaciers, and more recently the observed global temperature trend. This suggests that changes in the Sun's activity are affecting both the production of carbon-14 and the climate on earth. Solar variations are probably the driving force behind the temperature cycles revealed by the Camp Century core.
(Adapted from data supplied by John Eddy, US National Center for Atmospheric Research)

in the Earth's atmosphere (or something else I'm not clever enough to guess at) causes the Earth's climate to march in step with solar variations on these timescales. But whatever the relationship, there is no room to doubt that they do march in step. Although nobody has satisfactorily explained just how the link works, the evidence is too strong to be dismissed. And this, and other studies of how the Sun's activity varies from one cycle to the next, also shows that as well as the now familiar eleven-year solar cycle there are two longer term cycles. If the eleven-year cycle is likened to a wave running across the surface of the sea, one of these two cycles might be likened to a swell which moves the whole pattern of waves up and down, while the other is like the tide moving in and out. The overall pattern is complex, but predictable, so its component parts show up in the statistical tests. The lengths of those two additional solar cycles are eighty and 180 years, the same as the cycles that are found in the ice core record of temperatures.*

An astrogeological aside

Rocks deposited during an ice age in Australia 650 million years ago may be telling us something about the solar rhythms that influence climate on timescales of decades today. The rocks were formed from thin layers of sand and silt, deposited beneath a large stretch of water. Although they are no more than a few millimetres thick, these layers can be distinguished in the rocks from parts of South Australia today, and they bear a remarkable resemblance to tree rings. It is clear that each layer corresponds to the sediment laid down over a certain period of time, just as each tree ring corresponds to one year's growth. As yet, unfortunately, there is no firm means of distinguishing between two plausible explanations for the production of these deposits. One says that they are annual layers, and that their thickness each year was influenced by climatic factors sensitive to changes in the Sun's activity. The other suggests that the layers were produced daily, over an interval of about fifty-three

*The eleven-year cycle does *not* show up, presumably because the variation is too rapid to give the climate system time to respond. What we see in the proxy records, in effect, is the averaged-out influence of the strength of each solar cycle, the variations from decade to decade, rather than from year to year.

years, and that the thickness variations were caused by the changing tidal influence of the Moon. But even if the layers are a result of a lunar, not a solar, influence, in a curious scientific twist their discovery has stimulated research into the pattern of solar activity today, and that research leads to an unambiguous forecast of what we can expect in the 1990s and beyond.

The ancient sediments (so ancient that they were being laid down more than two hundred million years before life emerged from the sea onto the land) were discovered by George Williams, now of the University of Adelaide, who first drew attention to the possible link with sunspots in the early 1980s. He found the banded pattern in a section of sedimentary rock ten metres thick, from Pichi Richi Pass near the western end of the Flinders Ranges of eastern South Australia. The layers in the rock are each between 0.2 and three millimetres thick, and the thickness of the layers changes in a systematic way across the entire ten metres of the sample. Patterns of thick and thin layers follow a repeating cycle in sets of about twelve layers (or 'laminae') separated by darker bands. Some cycles contain as few as eight layers, others as many as fifteen, but the average length is twelve layers per cycle; this is, already, reminiscent of the pattern of solar variations, in which the length of an individual sunspot cycle may be anything from eight to fifteen years, but the average over the past century and a half is 11.2 years. Williams examined in detail one stretch of 1337 consecutive laminae, on the fine scale, and a run of 1580 consecutive cycles, marked by the darker bands, in a search for evidence of any regular fluctuations in the way the sediments had been deposited. These revealed rhythms very similar to present day solar rhythms, but with the addition of a modulation about 315 cycles long.

This led to the suggestion that the layers were indeed produced annually, and that they might be the ancient equivalent of lake deposits known as varves that form today. As we saw in Chapter Three, in lakes at high latitudes that are frozen over during the winter, silt and sand is carried down from the surrounding hills in a rush each spring during the thaw. This sedimentary material settles to the bottom of the lake and during winter, when the surface of the lake is frozen, the water clears almost completely. So each pulse of new spring material forms a distinctive layer on the lake floor.

If the sediments studied by Williams were formed in a similar fashion, they provide a record of thousands of years of variation in the run-off of water to a Precambrian ice age lake during the spring

83

thaw 650 million years ago. The similarity between the rhythms found in these rocks and solar rhythms today must, in that case, be a result of a strong solar influence on the climate of the frozen late Precambrian world, but nobody can say exactly how such a solar-terrestrial link would have operated. This is the weak point in the whole scenario. Astronomers, however, would have no trouble accepting either that the Sun's basic cycle used to be twelve years long and has slowly shifted to an 11.2-year pulsebeat, or that the thousands of years of data from the Precambrian actually provide a better guide than the few decades of modern data, and that the long-term average of the solar cycle is really closer to twelve years than to eleven years, even in modern times. They simply don't have enough good data to rule out either possibility.

Williams himself, though, had decided by the end of the 1980s that the rival lunar hypothesis was probably a better bet as an explanation of the variations in these sediments, which are sometimes known as the Elatina formation. It wasn't that he found any objection to the solar hypothesis, but rather that he found new evidence that similar sediments could be produced by tidal effects. He was persuaded by the discovery of other sedimentary rocks from South Australia, laid down at about the same time (give or take a couple of million years) as the Elatina formation, which contain thicker sedimentary stripes. These make up cycles containing fourteen or fifteen layers, with each of the laminar layers made up of a pair of 'semi-laminae'. This could represent the influence of the twice-daily lunar tide at work, repeating with a fortnightly rhythm.

In that case, the layers were probably formed from silt being carried down into a bay by a river. At low tide, the river rushes out strongly into the bay, carrying lots of large particles of suspended material with it. But at high tide the river is held back, almost as if by a dam, and only the finest silt filters through into the waters of the bay. Williams thinks that although the layers in the Elatina formation average only twelve per cycle, several thin layers may be obscured in many cycles by the dark bands which separate the cycles. He also says that each of the fine laminae could really be a pair of semi-laminae, together representing one day's deposits in such a bay.

If this suggestion is correct, the Elatina deposits are still a mine of information. They only cover a span of fifty-three years, not several thousand years, but they tell us exactly how the phases of the Moon varied in the late Precambrian, and also how many days there were in a month or a year back then. Williams infers that there were 30.5 days in the month (compared with 29.5 now), that there were 13.1 lunar months in the year (as against 12.4 today) and that there were four hundred days in the year, compared with the familiar 365. Researchers at the University of Arizona used the same data to calculate that the Moon was only 346,800 kilometres from the Earth 650 million years ago, and that it has therefore retreated at a rate of two centimetres per year out to its present distance of 384,400 kilometres. All of these figures are very much in the ball park that astronomers infer by calculating backwards from the present state of the Solar System and allowing for the way in which energy dissipated in tides has slowed the Earth's rotation and changed the Moon's orbit over geological time.

But the view that the Elatina laminae were produced by a solar influence still has its adherents, most notably Ronald Bracewell, of Stanford University.* He analysed the continuous run of 1337 laminae studied by Williams on the assumption that they represent 1337 years of climatic history. The usual statistical tests show that as well as the basic 'solar cycle' (but averaging about twelve years instead of roughly eleven today) the pattern is modulated by longer swells at both 314 years and 350 years. The 350-year variation seemed to cause the shift in the length of each individual cycle over the range from eight to fifteen years, while the 314-year cycle seemed to determine the size of the peak reached in any of the individual cycles.

All this would simply be playing with numbers, but Bracewell used the cycles he had discovered in the 1337 'year' long sedimentary record to calculate how solar activity ought to vary over the next few decades. Accurate records of sunspots only go back to the early 1800s, scarcely long enough to cover even half of the suggested 350-year cycle. There would be no hope of discovering

---

*It is a bizarre but seemingly genuine coincidence that two quite different 'explanations' can both account for the patterns found in the Elatina formation. In a further complication, which I have no intention of delving into here, some theorists argue that a combination of both solar and lunar influences might have been at work when these deposits were being laid down. A pessimist might argue that all this probably means that *both* explanations are wrong. An alternative possibility, which seems to have been ignored by the experts, is that the Elatina formation is indeed related to solar activity, but that the other sediments found by Williams, showing the clear pattern of fourteen or fifteen double-laminae, were produced by the tidal effect. The best argument in favour of the idea that the Elatina sediments are varves is still that we can see annual layers of sediment being formed in that way today. But even that is not conclusive.

the existence of cycles more than three hundred years long by analysing sunspot records covering less than two hundred years. But Bracewell found that if he used the cycles he had discovered in the Elatina sediments, and started out from a zero point at the solar minimum of activity that occurred in the summer of 1986, he could reproduce almost exactly the complex pattern of actual solar variations back to the year 1800. Though many people have tried since the solar cycle of activity was first recognized, nobody had previously found the right combination of short- and long-term cycles to reproduce the observed pattern so well. If the patterns in the Elatina formation were actually produced by lunar tides over a period of a few decades, not solar variations over thousands of years, this is a truly astonishing coincidence.

In a later extension of this work, Bracewell found that the basic twelve-year rhythm in the Elatina rocks might actually be produced by an interaction between the 314-year cycle and the genuine ancient counterpart of the present solar cycle of activity. The genuine cycle required to produce this effect would be 11.2 years long, *exactly* the same as the length of the solar cycle, on average, over the past century and a half. The argument is rather subtle, but you can see how this kind of effect works by making an analogy (imperfect, but easy to visualize) with the way the Earth orbits the Sun. Because there are 365 days in a year, and 360° in a circle, the Earth moves on in its orbit each day by almost one degree of arc. As a result, the rotation of the Earth from the time when the Sun is directly overhead (noon) to the time when the Sun is directly overhead again (noon the next day) is not 360° but almost 361°.

Over the course of a year, because the rotating Earth has traced a closed path around the Sun this adds up to precisely one extra turn of the Earth (and, of course, it always adds up to one extra turn, however many days there happen to be in a year; if there are fewer days, the effect is bigger each day, and if there are more days the effect is smaller each day). In other words, there is exactly a day's difference between the length of the year measured by observations of the Sun and the length measured by observations of distant stars.

If Bracewell's suggestion that something similar is happening to solar rhythms is correct, it means that there must be a core of magnetic field inside the Sun, rotating like an offset gyroscope, and spinning and precessing on the familiar timescales that show up from sunspot studies. Such an oblique magnetic rotator could be held to its steady period by an interaction involving the gravitational influence of Jupiter, the largest planet in the Solar System, which has an orbital period 11.86 of our years long. On that picture, the variations in the length of each sunspot cycle would simply be a result of variations in the time it takes for effects from the deep interior of the Sun to work their way out to the surface, and the internal 'clock' would have maintained its steady rhythm for hundreds of millions of years by feeding off the energy produced by nuclear fusion processes at the heart of the Sun.

But we can leave the debate on such issues to the astronomers. All this might seem like a far cry from the question of how the world's climate is likely to change in our lifetime, except that Bracewell's controversial claims also add weight to a forecast of temperature trends that was made in the mid-1970s, before anyone suspected that information about patterns of solar variations today might be extracted from Precambrian rocks.

A successful forecast?

Like all good scientific theories, Bracewell's makes predictions that can be tested. By running his calculations backwards in time from the middle of 1986, he gave a 'backcast' of how solar activity ought to have varied in the past, and found that this matched up closely to the pattern that had actually been observed. But this wasn't a real test of the theory, because he already knew what pattern of variations he was looking for. Hindsight is a wonderful aid to explaining anything. Running the same pattern of cyclic variations forwards from mid-1986, he predicted how solar activity ought to change in the years, decades and even centuries ahead, if the Elatina cycles are a reliable guide. So the first real test of his ideas is now taking place. The immediate forecast, published in 1986, was for a rise in solar activity up to 1991, with a relatively high peak of activity, above a hundred in the units used to measure sunspots.* Writing at the end of 1988, and comparing Bracewell's forecast with the official figures published each month by the Sunspot Index Data Centre in Brussels, the prediction seems to be bearing up rather well. Which just might be of interest to climatologist Wallace

---

*This is called the 'sunspot number', but refers to the area of the Sun's surface covered by dark spots, not to the actual number of individual spots counted on the solar disc.

Broecker, of the Lamont-Doherty Geological Observatory in New York.

If anyone can be regarded as the person who 'blew the whistle' on the anthropogenic greenhouse effect, Broecker was that man. In the early 1970s, he took a careful look at the implications for the global climate if human activities continued to pump carbon dioxide out into the air at an increasing rate. He used a simple 'business as usual' forecast of economic growth and associated industrial activity up to and beyond the year 2000, and he assumed that doubling the amount of carbon dioxide in the air would raise the average temperature of the world by 2.4°C. That figure came from what are now some rather venerable calculations of the greenhouse effect, a computer simulation run at Princeton University in 1967. (To put them in perspective, the computer used to run them was less powerful than the PC I am using to write these words.) In the early 1970s, there was, in fact, a wide variety of such estimates to choose from, some much bigger than this figure, others a lot smaller. Either by luck of judgement, though, Broecker picked one that turns out to be close to the figure that modern computer simulations (using machines far more powerful than my home PC) agree on. Broecker then made the completely unjustifiable guess that if doubling carbon dioxide concentration warms the world by 2.4°C, then each ten per cent rise in carbon dioxide concentration ought to warm the world by about 0.3°C. This was unjustified because at that time computer simulations could not say anything about such small changes in carbon dioxide, nor tell us how the climate would change on the way to a warmer world. Doubling the amount of carbon dioxide was the smallest change they could notice, and more subtle details are only just beginning to be filled in now, as we move into the 1990s. But Broecker was determined to get a forecast for the years up to the end of the century, before the additional carbon dioxide would have reached the concentrations the computers of the early 1970s could describe adequately. And he had one more piece of information to add to his scenario, a stroke of pure genius.

It is no good predicting the strength of the anthropogenic global warming if you don't know how the climate is going to change anyway, by natural causes. A warming of 0.3°C wouldn't show up at all if something else happened – a big volcanic eruption perhaps – to *cool* the globe by 0.5°C. So Broecker took the temperature cycles that had been discovered from isotope studies of the Camp Century core (the cycles which I believe are related to solar activity, but which are just as real whatever their actual origin) and used them as the basis of his forecast. The Camp Century cycles alone showed that the warming the world had experienced in the first part of the twentieth century was not attributable to the anthropogenic greenhouse effect, and should have been reversed by about 1950, plunging the world back into much cooler conditions, reminiscent of the nineteenth century, by the middle of the 1970s. A slight recovery would only take temperatures back to the level of the 1890s by the end of the present century, and this would be followed by a renewed decline.

But when Broecker added his calculation of the anthropogenic greenhouse effect to the Camp Century cycles (*Figure 4.4*), he found that the increasing strength of the global warming would drastically reduce the fall in temperature from 1950 to 1975, and that by the early 1980s the world would be in the grip of an increasingly rapid warming. In a paper published in *Science* in 1975, Broecker caused something of a scientific sensation with his claim that 'the exponential rise in the atmospheric carbon dioxide content will tend to become a significant factor and by early in the next century will have driven the mean planetary temperature beyond the limits experienced during the last thousand years'.

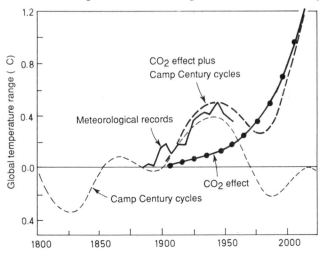

Figure 4.4
By combining the Camp Century cycles with calculations of the carbon dioxide greenhouse effect, Wallace Broecker warned in 1975 that the world was on the brink of a rapid global warming. Temperature variations since 1975 closely match his forecast.
(Source: Science, volume 189 page 461, 1975)

Outside those scientific circles, Broecker's words made very little impact in a world where there was no sign of rising temperatures. The natural climate cycles, holding the warming at bay for three decades from 1950 to 1980, may have lulled politicians and planners into a false sense of security. Everything that has happened since 1980 is very much in line with Broecker's forecast, and with the natural cycles now adding to the greenhouse effect until about the year 2020 it seems clear that we can expect an even more rapid rise in temperatures over the next three decades than could be produced by the anthropogenic greenhouse effect alone.

Viewed against the background of climatic changes since the latest ice age, which shows that for most of the past thousand years the world has been colder than during any other millennium of the present interglacial, you might think that this extra greenhouse effect is a good thing, returning us to conditions in which civilizations flourished in the little optimum. Taking a longer perspective, anything which might delay the end of the present interglacial, and hold back the advance of the next ice age, is 'obviously' a good thing from the point of view of anyone who lives in North America, Europe, Australia or the Soviet Union. But Broecker's simple projection, made more than ten years ago on the basis of only sketchy scientific information, highlights the real problem of the greenhouse effect. A *little* warming might indeed be a good thing. But through the increasing release of carbon dioxide and other greenhouse gases into the atmosphere, we are faced with a very rapid, very large increase in global temperatures. This is almost certainly going to be too much of a good thing – and we can see why by taking a look at just how human activities are now interfering with the natural cycles of carbon that have been part of the workings of Gaia for scores of millions of years.

FIVE

Natural and Unnatural Carbon Cycles

After the pioneering work by Svante Arrhenius and Thomas Chamberlain at the end of the nineteenth century, surprisingly little scientific attention was paid to the carbon dioxide greenhouse effect for more than fifty years. In the first half of the twentieth century, the day when human activities might lead to a doubling or tripling of the 'natural' concentration of carbon dioxide in the air still seemed a long way off, and the few researchers who discussed the issue in the decades between the two World Wars seem to have been concerned not by the climatic consequences of a build-up of carbon dioxide in the air, but by the fact that valuable reserves of fossil fuels, that could never be replaced on any human timescale, were being squandered. But two things happened in 1957 which make that year stand out, with hindsight, as the beginning of the modern era of interest in the carbon dioxide greenhouse effect.

The first landmark event was the publication of an article in the journal *Tellus* by Roger Revelle and Hans Suess, of the Scripps Institution of Oceanography. They discussed how carbon dioxide from the atmosphere is taken up by the oceans, suggested that this is such a slow process that carbon dioxide added to the air by human activities would stay there for centuries, and coined a phrase which has now been repeated (usually without attribution) so many times that it has become a cliché. Pointing out that nobody at that time could predict the consequences of a build-up of carbon dioxide in terms of effects on climatic patterns and living things, they said that 'human beings are now carrying out a large-scale geophysical experiment', testing the consequences of the green-house effect by allowing the concentration of carbon dioxide in the atmosphere to increase. In other words, we are experimenting on ourselves, without knowing in advance how the experiment will turn out. From time to time over the past thirty years – most recently during Congressional hearings on the greenhouse effect in 1988 – that phrase, or a variation on it, has surfaced and made headlines. It is as accurate a description of what is going on now as it was in 1957.

The other important event that happened at about the same time was that scientists began to monitor the amount of carbon dioxide in the air on a continuous basis. Measurements had been made before, from time to time and place to place, but not in any really systematic fashion. During 1956 and 1957, however, geophysicists from around the world carried out many studies of our planet as part of what was known as the International Geophysical Year; the first artificial Earth satellites were launched as part of this programme of research. Another scientist from the Scripps Institution, Charles Keeling, made his contribution to the global research effort by starting a series of measurements of atmospheric carbon dioxide – measurements that have continued almost without a break right up to the present day. The measurements are made at the Mauna Loa Observatory, on the top of a mountain in Hawaii, far from any sources of industrial pollution; a second series of measurements has been carried out from the South Pole, with a break of about a year in 1963. Together, these measurements (and other series of observations started more recently) show the build-up of carbon dioxide in the atmosphere. They also show an annual variation in the amount of carbon dioxide in the air, a seasonal change linked with the growth of plant life in the northern spring and summer (taking carbon dioxide out of the air) and its decay in the autumn and winter. In 1958, the average concentration of carbon dioxide in the atmosphere, recorded at Mauna Loa, was about 315 parts per million (ppm) by volume. In 1988, the annual average concentration passed 350 ppm (*Figure 5.1*).

Even today, only 0.035 per cent of the atmosphere is carbon dioxide. It is a small concentration indeed, but it plays a crucial role in keeping the Earth warm. And it is precisely because there is so little carbon dioxide in the air to start with that the amount added by human activities is a significant contribution – since 1958 alone, we have added, on these figures, thirty-five ppm to a concentration in that year of 315 ppm. In other words, over the past thirty years human activities have increased the carbon dioxide concentration of the atmosphere by just over eleven per cent of the concentration it had in 1958. But how much of the 315 ppm of carbon dioxide

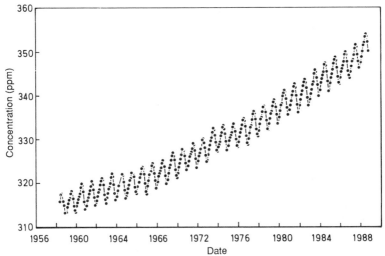

Figure 5.1
The Mauna Loa record of atmospheric carbon dioxide concentrations, measured in parts per million by volume.
(Source: Charles Keeling/Carbon Dioxide Information Analysis Center)

in the air at the end of the 1950s had itself got there as a result of human activities?

One of the clearest indications that the Mauna Loa record is, indeed, showing the effect of both natural cycles and human activities ('unnatural' influences) on carbon dioxide has come quite recently, from the study of air bubbles trapped in the ice of the Vostok core. Although the build-up of carbon dioxide in the atmosphere was clear from the Mauna Loa measurements before the end of the 1960s, at that time nobody could say for certain that this was solely a result of human activities, burning coal and oil and cutting down forests. For all anybody knew at that time, there might be natural processes which caused the carbon dioxide concentration to fluctuate by ten per cent, or more, over timescales of decades and centuries, just as there are natural processes which cause much smaller seasonal fluctuations each year. It might just be a coincidence that the carbon dioxide concentration was increasing at a time when people were digging up and burning a lot of fossil fuel. But the ice core data knocked that argument on the head. It shows that the concentration of carbon dioxide in the atmosphere has now reached levels higher than at any time in the past 160,000 years, as far back as the Vostok core can reach. There is more carbon dioxide in the air now, not just than there has been at any time during the present interglacial, but than there was at any time

93

in the *previous* interglacial. The peak carbon dioxide concentration shown in the Vostok bubbles is from 135,000 years ago when the concentration only reached three hundred ppm – less even than the concentration in 1958 when modern monitoring began. As for the present interglacial, although there was an increase in the carbon dioxide concentration of about ten ppm between 1200 and 1300AD, this stands out as an unusual – indeed, unique – natural fluctuation. Even so, it did not take the concentration as high as three hundred ppm. The present 'fluctuation' is far bigger, has happened more quickly, and can only have been caused by the interference of human activities with the natural carbon cycles. We can only estimate how big this anthropogenic influence will become in our lifetimes, and those of our children, if we have some idea of how those natural cycles work; the key role is played by the oceans, but we can only understand that role if we know how much carbon there is, stored in different 'reservoirs', that might be available to play a part in the carbon dioxide cycles.

Carbon cycles and reservoirs

Carbon is the most important chemical element for the existence of life on Earth. Of course, many other elements are essential for life as we know it. But carbon atoms have a unique ability to form complex compounds, including chains, rings and helices, which also contain atoms of many other elements, and this is the key to life.* Carbon chemistry is, indeed, so intimately connected with life processes that the study of the reactions that involve carbon is known as 'organic' chemistry. The study of all other chemical reactions and compounds is known as 'inorganic' chemistry. More than a million carbon compounds are known, and thousands of these are essential for life. Because of its ability to form so many different compounds, and because of its key role in living processes, carbon is always being moved around the biosphere. Carbon compounds are continually being created, altered into other compounds, and destroyed, on timescales that range from a few seconds to many millions of years; but as far as the anthropogenic greenhouse effect is concerned we are only interested in how carbon moves around on timescales of about a century or less, and from now on I shall ignore the long,

*The reasons for this are explained in my book *In Search of the Double Helix* (Bantam/Corgi 1985).

94

slow processes that were discussed in Chapter Two. The 'carbon cycle', as it is called, is one of the most impressive and ubiquitous features of Gaia – just as the blood circulating in your body links every cell with every other part of the body, so the carbon circulating through Gaia links all her components into one web. And just as blood (which, of course, is made from carbon compounds) carries oxygen from the outside environment into your body, and carries carbon dioxide out to be expelled into the physical world outside, so the flow of carbon through the biosphere also involves interactions with the physical environment, the part of the world that we are used to thinking of as non-living, but which Jim Lovelock tells us is an integral part of Gaia.

The key link between the biosphere and the physical world is the process of photosynthesis, where plants on land and in the sea take carbon dioxide out of the air and, with the aid of sunlight, use this and water (or, in some cases, another source of hydrogen) to produce molecules of formaldehyde and oxygen (or an equivalent by-product). The formaldehyde produced by photosynthesis is the basis of all the molecules of the familiar variety of living things. Once carbon is 'fixed' in this way, it is available for the thousands of biochemical reactions that build up fats, proteins, DNA and the rest of life's complexity; any living things that are incapable of photosynthesis (such as ourselves) are completely dependent on obtaining supplies of biological material (food) by eating other things that *are* capable of photosynthesis (plants), or by eating creatures that have themselves fed on plants.* Without photosynthesis, we would not be here.

Even plants, however, cannot exist by photosynthesis alone. Photosynthesis is a process that stores energy from sunlight in a chemical form. This energy can be released when the organic compounds built up by photosynthesis are allowed to react with oxygen from the air – a slow form of burning. Then carbon dioxide is released to the atmosphere, and the energy that is liberated is used in a further series of chemical reactions that build up different organic molecules. This is how plants grow, and it is the way that animals, including ourselves, get their energy (which is why we breathe in oxygen and breathe out carbon dioxide). The

---

*There are, in fact, some life forms that exist in the deep ocean and use energy from the heat of underwater geological activity, rather than sunlight, to build up the basic molecules of life. Interesting though they are, they are not relevant to the present story, and I shall ignore them.

process which takes oxygen out of the air and combines it with carbon from organic molecules in living things is called respiration, and it occurs in both plants and animals. Respiration in plants goes on all the time, but photosynthesis only goes on during daylight hours, when energy is available from the Sun. So, on balance, plants release oxygen during the day and carbon dioxide at night.

The impressive effects of all this on the carbon dioxide content of the air can be measured in a forest. In the morning, photosynthesis proceeds apace, but there is only a slow rate of respiration. By midday, the amount of carbon dioxide in the air between the treetops is as much as fifteen ppm below the average for a whole twenty-four-hour period; but in the afternoon, when the air is warmer and more humid than in the morning, respiration picks up and carbon dioxide concentrations stop falling. During the night, photosynthesis stops, but respiration continues, so the carbon dioxide concentration in the forest air increases. The carbon dioxide tends to sink to the ground during the cool of the night, because there are no convection currents to carry it upward when there is no sunshine heating the ground. There is also a contribution of carbon dioxide from the decay of rotting organic material in the soil (another form of slow burning). The combined effects push the nighttime carbon dioxide concentration of the air at ground level in the forest well above four hundred ppm.

In some parts of the world, where forests are growing rapidly, the net effect of all this activity is to take carbon dioxide out of the air and store it in living wood. Tropical rainforest can fix between one and two kilograms of carbon in this way for each square metre of land area each year. This sounds a modest figure, but it is roughly the same as the amount of carbon (in the form of carbon dioxide) in a column of air one square metre in area extending from the surface of the Earth all the way up through the atmosphere. This is the highest rate of carbon fixation anywhere on the land surface of the Earth. Agricultural activity outside the tropics, in regions such as Europe and North America, may fix no more than four hundred grams of carbon per square metre each year (carbon which, of course, gets eaten and turned back into carbon dioxide by respiration).

Assessments of how much carbon is fixed by land plants each year altogether vary enormously, but Bert Bolin, of the University of Stockholm, has estimated that a fair guess would be twenty to thirty billion metric tons (tonnes) of carbon each year. The amount of carbon taken up by phytoplankton in the oceans is even harder to estimate, but it was until recently fairly widely accepted that this might amount to about forty billion tonnes per year, rather more than the amount being fixed by land plants. Some recent studies, however, have suggested that phytoplankton may be far more active in this way than has previously been appreciated – more of this later. Whatever the total figure may be, until human interference began to upset the natural cycles, the amount of carbon being fixed each year was almost exactly balanced by the amount being released, in the form of carbon dioxide, from other processes, such as respiration and the decay of dead plants and animals. This is clearly established by the ice core records, which show that until the nineteenth century the carbon dioxide concentration of the atmosphere had never differed, during the entire present interglacial, by more than ten ppm from an average value of 280 ppm. There is, though, a crucial difference between the activity of phytoplankton and the activity of land plants. Although land plants take carbon dioxide directly from the air, and release oxygen into the air, phytoplankton use carbon dioxide that is dissolved in sea water, and the oxygen they release also dissolves in the sea. This oxygen is itself utilized by sea animals in respiration, and ultimately gets back into the sea again as dissolved carbon dioxide. There is a closed system of carbon circulation around the organisms that inhabit the ocean. But as the carbon dioxide concentration of the atmosphere increases, more carbon dioxide dissolves in the oceans and is available to the phytoplankton. One of the most important problems in the whole of the investigation of the anthropogenic greenhouse effect is how much of the carbon dioxide we are putting into the atmosphere will be taken up by the oceans, and how much will remain in the air to contribute to the growing greenhouse effect.

Because phytoplankton are not directly linked to the carbon dioxide reservoir in the atmosphere, seasonal changes in their activity do not show up strongly as an influence on the carbon dioxide concentration of the atmosphere. The pronounced seasonal effect that can be seen in the measurements made at Mauna Loa is primarily a result of the changing activity of plants on land. Activity in the tropics does not vary much over the course of the year, but seasonal effects are more pronounced at higher latitudes. Since most of the Earth's land surface, outside the tropics, is in the northern hemisphere, that means that the cycle follows the northern seasons. Between April and September, the atmosphere north of 30°N loses nearly three per cent of its carbon dioxide, the equivalent of about four billion tonnes of carbon. Decay processes in the soil go on all the time, releasing perhaps half this amount of carbon dioxide into the air over that six months, so the actual amount of carbon dioxide being fixed at high northern latitudes in summer must be at least six billion tonnes each year.*

Estimates of how much carbon is actually stored in the wood of forests today have varied enormously over the years. By and large, the figures have been revised downwards as more accurate techniques for measuring the amount of carbon have been developed, and as larger areas of the Earth's surface have been surveyed (and as tropical forests, in particular, have been destroyed in recent years). Tropical forests today probably account for only about 170 billion tonnes of carbon, out of a total of some 620 billion tonnes in the form of 'standing biomass' – trees and other large plants. But there is probably about twice as much carbon, at least 1300 billion tonnes, stored in organic material in the soil, including peat. The atmosphere itself contains about 725 billion tonnes of carbon in the form of carbon dioxide. (When a tonne of carbon is burnt, it produces about four tonnes of carbon dioxide, as each atom of carbon, with a weight of twelve atomic units, combines with two atoms of oxygen, each with a weight of sixteen atomic units, to give a molecule with a weight of forty-six units. So 'a tonne of carbon in the form of carbon dioxide' is the same thing, roughly speaking, as 'four tonnes of carbon dioxide'.)

Fossil reserves of carbon, representing the remains of long dead organisms that, ultimately, got their carbon from photosynthesis, are much bigger than any of these figures. Estimates of reserves of oil and gas are little more than educated guesswork, and tend to change each time a new oilfield is discovered, but such reserves probably contain no more than a few hundred million tonnes of carbon. If all of this were extracted and burnt, it would do no more than increase the present day concentration of carbon dioxide in the atmosphere by fifty per cent. The amount of carbon stored near the surface of the Earth's crust in the form of coal, however, is at least ten thousand billion tonnes, and sixty per cent of this, equivalent to nearly ten times the amount of carbon in the atmosphere in the form of carbon dioxide today, could be recovered and burnt using feasible mining technology. This is what makes it possible

*These figures come from an article by Bolin in *Scientific American*, September 1970. They should probably be increased slightly today, since the Mauna Loa record shows that the seasonal fluctuation has got slightly bigger as the amount of carbon dioxide in the air has increased.

for human activities, in the short term, to overwhelm completely the natural concentration of carbon dioxide in the atmosphere.* But even these huge reserves of carbon are more than matched by the amount stored in the sea.

Oceanic cycles

Just as the amount of carbon dioxide present in the entire atmosphere can be estimated by measuring the concentration of carbon dioxide in a small sample of air and multiplying by the volume of the entire atmosphere, so the amount of carbon dioxide stored in different layers of the ocean can be estimated by measuring the concentration in samples of sea water from different depths and different regions of the globe. But the layering of water in the seas is different from the layering of the atmosphere, and this affects the way ocean currents circulate. Because the Earth's surface is warmed by the heat of the Sun, the bottom of the atmosphere is warmer than the air just above. This hot air rises, and produces convection currents which drive the large scale circulation of the atmosphere. But the Sun warms the surface of the ocean, as well as the land, and so the *top* layer of the sea is warmer than the waters in the depths below. The cold water cannot rise up into the warm water above, and stays in the deeps for a long time. Deep, cold ocean water only reaches the surface in significant amounts near the poles, where the surface water is also cold. Over most of the surface of the sea, for most of the time, there is very little interaction between the relatively warm surface layers, which exchange carbon dioxide with the atmosphere and in which most sea life exists, and the cold, deep layers, which, among other things, store carbon dioxide out of contact with the atmosphere for up to a thousand years.

*In the *long* term (millions of years), Gaian processes along the lines discussed in Chapter Two will return the carbon dioxide concentration of the atmosphere back to the levels that prevailed before human interference. A hundred thousand times as much carbon dioxide as there is in the atmosphere today is stored in sedimentary rocks such as limestone and dolomite. This is almost exactly the same as the amount of carbon dioxide in the atmosphere of Venus, where no similar process of sedimentation has occurred, and where a runaway greenhouse effect operates. Given millions of years, the Earth – Gaia – will safely lock away any fossil carbon that human activities release. The problem we face is that of how the climate might change on timescales of decades and centuries, long before those slow but sure processes can restore the long-term equilibrium.

When carbon dioxide dissolves in the oceans, it reacts with water to form carbonates. Carbon in this form is often referred to as 'dissolved inorganic carbon', or DIC, to distinguish it from the dissolved remains of living things, known as dissolved organic carbon, or DOC. Altogether, the oceans contain about fifty times as much dissolved inorganic carbon (about thirty-eight thousand billion tonnes) as the amount of carbon stored as carbon dioxide in the air. Measurements of the amount of radioactive carbon-14 in the air and in the sea show that the average lifetime of a molecule of carbon dioxide in the air, before it is transferred to the sea, is about eight and a half years. Every year, about a hundred billion tonnes of atmospheric carbon dioxide is dissolved in the sea, and is replaced by almost exactly the same amount of carbon dioxide being released from the sea. The rate at which these interchanges take place depends on the amount of carbon dioxide in the atmosphere, the amount of dissolved inorganic carbon in the surface layer of the oceans, and the temperature. When the ocean is *colder*, it can absorb *more* carbon dioxide; there is a disturbing, but real, prospect that as the world begins to warm because of the anthropogenic greenhouse effect this oceanic balance will tilt and more carbon dioxide will be released to the atmosphere, strengthening the greenhouse effect in a positive feedback process.

Estimates of the amount of dissolved organic carbon in the oceans are very poor, and based on only a few measurements. A widely accepted rule of thumb is a figure of a thousand billion tonnes, but as we shall see later there is some evidence that this figure should be revised upwards. In the long term, the DOC is very important. It is produced by biological activity, as phytoplankton fix carbon dioxide and are eaten by zooplankton, which are eaten by larger animals, and so on up the food chain. A small fraction of the organic material escapes from the chain each year, and settles to the ocean depths where, as we saw in Chapter Two, it becomes part of the long term cycles that maintain the temperature balance of the Earth. The concentration of DOC is a factor in determining the strength of these cycles, which, as we have seen, may also be involved in regulating the temperature of Gaia.

There is intense competition for food among the organisms that live in the sea, and as a result most of the surface layers today contain only modest amounts of nutrients – the rest have been eaten. The exceptions are in regions where cold water from the depths wells up to the surface, carrying nutrients with it; so the most productive regions of the ocean, in terms of biological activity, are where cold water reaches the surface around Antarctica. Even there, though, biological activity is limited. There are high concentrations of both nitrogen and phosphorus in the water, and these are elements that living things need. If they are not being taken up by living things, that must be because those living things also need something else which is in short supply. That 'something else' might be light, or it might be iron (or both); in either case, it is possible that it provides a link between the living components of Gaia and the astronomical cycles of the Milankovitch climate rhythms.

As well as being colder deeper below the surface, deep ocean water is more salty than the water above, which makes it more dense and helps to keep it in place, below the relatively thin layer that comes into contact with the atmosphere. This surface layer, which gets mixed up by the action of wind and waves, is only about seventy-five metres deep, on average. It is the concentration of dissolved inorganic carbon in this layer that determines the concentration of carbon dioxide in the air above, provided the system is given time to come into equilibrium. For the surface layer alone, this takes only about ten years (at most a few decades), the time needed for a complete interchange of all the atmospheric carbon dioxide with dissolved carbon dioxide; but even over a period of ten years, human activities will have changed the carbon dioxide concentration of the atmosphere, so today even this rapid carbon cycle never reaches equilibrium. The cold, deep water contains a higher concentration of DIC than the warmer, surface layer, and if this extra carbon dioxide were spread evenly through the entire ocean then the concentration of DIC at the surface would be about fifteen per cent higher than it is today. Some of this extra carbon dioxide would leak out into the atmosphere, raising the concentration of carbon dioxide there to at least seven hundred ppm, twice the present day figure.

The deep ocean waters originate in wintertime in the Norwegian and Greenland seas in the northern hemisphere, and in the Weddell Sea in the Antarctic. The surface waters become so cold in wintertime that they sink downwards, and flow as deep ocean currents towards lower latitudes. The deep currents flow into and around the great ocean basins of the world, and are eventually pushed upward into shallower waters, and return to the surface. This happens, for example, along the coast of Chile, where the rising water, rich in nutrients, causes an explosion of biological activity that produces great shoals of fish. This and similar currents

101

play an important part in determining the weather patterns of the world from year to year, as we shall see in Chapter Seven.

Cold water from the Norwegian and Greenland seas stays deep in the Atlantic Ocean for about three hundred years (the 'residence time' is estimated from measurements of current flow and from the ubiquitous radioactive isotopes). In the south, Antarctic bottom water stays deep below the Indian Ocean for about 335 years, and beneath the Pacific for at least six hundred years before being pushed up to the surface, in effect, by the pressure of more cold water behind it. In round terms, it takes a thousand years for surface water that descends at high latitudes to complete its journey and 'mix' right through the ocean. The water that is returning to the surface today carries with it a memory, in the form of the concentration of dissolved inorganic carbon, of the carbon dioxide concentration in the atmosphere hundreds of years ago; the changes in carbon dioxide concentration produced by human activities will not have affected the whole of the deep ocean until almost the year 3000.

All of this makes it particularly hard to estimate how much carbon dioxide will remain in the atmosphere over the next hundred years or so, because we are not dealing with equilibrium conditions. At present, a great deal of the carbon dioxide being released by human activities is dissolving in the cold waters at high latitudes in winter and being carried away into the depths. It is not clear how long this process can continue. If the world warms significantly, and winters are no longer so cold at high latitudes, then the process by which this cold bottom water forms may be slowed down, and that might mean more carbon dioxide stays in the air (changes in oceanic circulation would also affect the climate directly, as we shall see later).

In spite of the uncertainties and difficulties, researchers have tried to calculate, using computer simulations of the ocean circulation and the interaction of surface waters with carbon dioxide, the rate at which excess carbon dioxide produced by human activities is being taken up by the sea. The best of these computer 'models' include interactions between the atmosphere and the mixed layer over most of the oceans, plus an interaction between the atmosphere and a 'polar outcrop' where the deep layer stretches up to the surface at high latitudes. The calculations suggest that about half of the carbon dioxide being produced by human activities is being taken up by the oceans today. This is broadly in line with measurements of how rapidly the concentration of carbon dioxide

102

is building up in the atmosphere, compared with estimates of how much carbon dioxide is being released by the burning of fossil fuels. But the calculations also show that the oceans can absorb additional carbon dioxide more efficiently if there is a slow, long-term release of the gas by human activities; the more rapid the release of carbon dioxide is, the bigger the fraction that stays in the air in the short term. This is crucial in estimating how much carbon dioxide will stay in the atmosphere in the decades ahead, as we shall see in the next chapter; it is also helpful in calculating how much carbon dioxide must have been produced by human activities over the past couple of hundred years in order to raise the concentration in the atmosphere to the 350 ppm measured today. The other key to understanding the build-up of carbon dioxide in the past comes from more isotope studies.

An isotopic cornucopia

Bubbles of air from Antarctic ice cores contain a record of how the carbon dioxide content of the atmosphere has changed in recent years, on a fine timescale as well as over many millennia. These air bubbles can even tell us how much of the extra carbon dioxide put into the atmosphere by human activities over the past couple of hundred years comes from burning fossil fuel (mainly coal), and how much comes from the destruction of forests and the release of organic carbon from the soil as natural ecosystems are disturbed. The evidence is trapped in the bubbles in the form of isotopes of carbon.

There are actually seven isotopes of carbon – atoms which have the same chemical properties but different weights – that can be manufactured with reasonable ease in atomic reactions. But we have already met the only three that are important in the natural environment of the Earth. Two of these isotopes are stable, and differ from each other by one atomic mass unit. Carbon-12 is by far the most common isotope, and about 98.9 per cent of the natural carbon in the world is in this form. About 1.1 per cent is in the form of the heavier stable isotope, carbon-13; only a small fraction of one per cent of the carbon in the environment is in the form of the radioactive isotope, carbon-14, that is produced in the atmosphere when cosmic rays interact with atoms of nitrogen. Carbon-14 has a 'half life' of 5,730 years. In any collection of carbon-14 atoms, exactly half of them 'decay' (converting back into stable nitrogen) in that time. Then, half of the remaining carbon-14 decays in the next 5,730 years, and so on. There is a rough balance between the amount of new carbon-14 being manufactured each year and the amount decaying in this way; that balance maintains a pool of about eighty tonnes of carbon-14 in the environment of the Earth – but just as the level of water in a pond may stay roughly the same even though water is running in one end and out the other, the actual carbon-14 atoms in that pool of radiocarbon are constantly changing. Because the carbon-14 (radiocarbon) is manufactured in the air, but is constantly being removed by this process, a comparison of the proportion of carbon-14 (compared with the total carbon content) of the atmosphere with that of the ocean reveals how long it takes for carbon atoms, in the form of carbon dioxide, to be taken up by the water – the slower the uptake, the less carbon-14 there will be by the time the carbon dioxide has dissolved. But this is only the beginning of the isotope story.

Although the three isotopes react in the same way chemically, they do not react at the same rates. Heavier isotopes react more slowly, and are taken up more reluctantly by living things. This reluctance is not a vague effect, but as precise and quantifiable as the half life of carbon-14. From a known atmospheric mixture of the different isotopes, chemical processes, including biochemical processes, will always select out a definite different mixture of the isotopes. In living things, the proportion of carbon-13 to carbon-12 is less than in the air; and because the difference in mass between carbon-14 and carbon-12 is exactly twice the mass difference between carbon-13 and carbon-12, the effect is twice as strong, atom for atom, in the case of carbon-14.

The effects show up throughout the Earth's carbon reservoirs. In the ocean, for example, the proportion of carbon-13 in the dissolved inorganic carbon of the surface waters is about 1.5 parts per thousand *higher* than in the deep ocean, because living organisms in the surface layers prefer to take up carbon-12. In plants that live in the water, the proportion of carbon-13 is, by contrast, about twenty parts per thousand *lower* than in the dissolved inorganic carbon of the waters in which they live. Similarly, land plants, including trees, contain less carbon-13 relative to the amount of carbon-12 they contain, than the carbon dioxide of the atmosphere. Fossil fuel, such as coal, is made from the remains of dead plants, and also contain less carbon-13 than the air – indeed, the ratio of carbon-13 to carbon-12 in coal will be the same as the ratio in the material of the living plants from which it was made. But living trees also contain carbon-14, which they have taken up from the air around them. Fossil fuel, by contrast, is millions of years old, so old that virtually all of the radiocarbon that was once in it has decayed. And this provides a crucial way in which to distinguish between carbon dioxide that comes from burning trees and carbon dioxide that comes from burning coal. The proportion of carbon-13 in fossil fuel, compared with carbon-12, is about eighteen parts per thousand lower than the proportion in the atmosphere today. When fossil fuel is burnt, the overall concentration of carbon dioxide in the air increases, but the proportion of carbon-14 *decreases*, since the 'new' carbon dioxide contains no radiocarbon. At the same time, the proportion of carbon-13 *decreases*, because the fossil fuel contains proportionately less carbon-13 than the atmosphere does. When forests are cut down and carbon dioxide is released from destruction of the biomass, however, the resulting increase in the carbon dioxide concentration of the atmosphere is accompanied only by a decrease in the proportion of carbon-13; the carbon-14 is not affected in the way that it is when fossil fuel burns.*

Vast amounts of carbon-14 were produced in nuclear weapons tests in the atmosphere in the 1950s and early 1960s, and this radiocarbon has disturbed the pattern of isotope ratios that existed before then. Researchers have been able to use this 'pulse' of radiocarbon as a tracer, watching its spread through the environment and using it, for example, to aid their calculations of how long it takes for all the carbon dioxide in the atmosphere to be interchanged with carbon dioxide in the oceans. But this distortion of the isotope balance means that measurements of atmospheric carbon-14 made since 1950 are not a straightforward guide to how much fossil fuel is being converted into carbon dioxide each year. That doesn't matter too much, since we have good records of how much fuel has been burnt since 1950; climatologists are much more interested in finding out what was going on in the decades before that time, and the isotope technique gives them a way to tackle the problem.

The initial attempts at unravelling the isotope record, though,

*In fact, since living things prefer to take up lighter isotopes, there is slightly less carbon-14 in the living material, and even in a tree a hundred years old a tiny amount of this will have decayed. So there is in principle a very small dilution of the carbon-14 content of the atmosphere even when living biomass is destroyed. Compared with the dilution that occurs when fossil fuel is burnt, however, this effect is so small that it can be ignored.

did not use air bubbles from ice cores. Another way to find out what was going on in the atmosphere in the years before 1950 is to look at the carbon content of tree rings, layers of wood which are each laid down in a single year, and which can be dated precisely by counting the rings back from the present day. In some ways, this is easier than studying ice core samples: for a start, trees grow in more convenient locations, and you don't have to go to Antarctica and drill a core for analysis – and the tree ring technique was developed first.

Tree rings and ice cores

The amount of carbon-14 produced in the atmosphere each year is just under ten kilograms, with roughly the same amount decaying, somewhere on Earth, in the same time. The balance is only roughly constant because the amount of radiocarbon being produced varies – for example, when the Sun's activity varies. With so little carbon-14 around at all, the amount that is present in a living organism, such as yourself, is tiny – 0.3×10^{-12} (a decimal point followed by twelve zeros and a 3) of a gram for every gram of carbon in the organism.* But because the carbon-14 is radioactive, and emits an electron when it decays into nitrogen, it can be detected (essentially by counting the electrons produced) and measured even in such minute quantities. When tree rings are analysed in this way, as they were in the early 1980s, they show clear signs of the decline in carbon-14 concentration since the late nineteenth century that was predicted on the basis of fossil fuel emissions. The decline, known as the Suess effect (after Hans Suess, who first described it in 1955), amounts to a change in the proportion of carbon-14 in the atmosphere (compared with total carbon) of about twenty-five parts per thousand (2.5 per cent) between 1850 and 1950. Before 1850, the proportion was roughly constant.

The tree ring measurements also show a similar decline in the

*To put that in perspective, a gram of carbon contains about 5×10^{22} atoms. So, even the 0.3×10^{-12} of that that is carbon-14 still accounts for about 15×10^9 atoms – fifteen *billion* atoms of radioactive carbon out of every gram of carbon in your body. The chance of a single one of those atoms decaying in any one year is ½ in 5,730, just under one in twelve thousand. So for each gram of carbon contained in you, more than a million of these radioactive atoms decay inside your body every year – 2,500 every day, a hundred every hour, almost two every minute. And you never feel a thing!

106

proportion of carbon-13 in atmospheric carbon dioxide over the same period. But this effect is even smaller than the carbon-14 Suess effect, because there is more carbon-13 around in the atmosphere to be diluted. The influence of the depletion of carbon-13 on atmospheric carbon dioxide when coal or wood burns is less than the effect on the proportion of carbon-14 in the air when coal or oil burns, and the change in carbon-13 amounts to less than two parts per thousand over the same interval, from 1850 to 1950. This change is much less than the change that occurs when carbon dioxide is taken up by plants, which alters the fractional amount of carbon-13 by seven parts per thousand, or the fractionation (about twenty-five parts per thousand) that occurs during the chemical reactions that form cellulose out of simpler organic compounds. Although researchers in the early 1980s were happy that the overall trend of carbon-13 in tree rings confirmed their expectations, nobody put very much faith in the detailed pattern of variations from decade to decade, because it seemed very likely that changes in the wood and environmental influences could be producing alterations after the rings had been laid down but while the trees were still alive. For a few years, there was a lively debate about how best to interpret the tree ring data. But we don't have to go into the details here, because in 1986 researchers from the University of Bern, in Switzerland, reported the first detailed analysis of the carbon isotope content of air bubbles trapped in the ice of a core drilled from Siple Station in Antarctica. This particular core comes from a region where the layers of ice laid down in recent decades can be identified, and the air bubbles they contain analysed to provide a decade by decade – almost a year by year – record of how the proportion of carbon-13 has varied over the past two centuries (and before).

This is now the best available record of such variations, and it shows that although the carbon-13 fraction was almost the same in 1750 as it was from the thirteenth century to the sixteenth century, between the eighteenth century and 1980 the fraction decreased by a little over one part per thousand. The crucial new information that this ice core study provides is that the decline in the proportion of carbon-13 in the atmosphere set in long before the decline in the proportion of carbon-14. Direct measurements of the overall concentration of carbon dioxide in air bubbles from other ice cores show that this concentration had already started to increase before 1800, when it was about 280 ppm (perhaps a little lower). Together, these measurements show that the build-up of carbon dioxide has been going on for longer than most researchers had previously suspected, and that initially the build-up was almost entirely caused by the release of carbon dioxide from the destruction of forests and the conversion of land use to agriculture. It was not until well into the twentieth century that the burning of fossil fuel became the major source of anthropogenic carbon dioxide (*Figure 5.2*). Two of the researchers involved in the original analysis of the Siple core, Hans Oeschger and Ulrich Siegenthaler, have combined all this information with the best available computer models of how carbon dioxide cycles between the air and the sea, and have inferred how human activities have altered the natural carbon dioxide balance, and where the 'unnatural' carbon dioxide comes from and goes to.

Unnatural carbon dioxide

The Bern researchers' work was summarized at a conference on 'The Changing Atmosphere' held in Berlin under the auspices of the Dahlem foundation in November 1987. We now know, from ice core studies, that the natural concentration of carbon dioxide in the air before human interference was no more than 280 ppm. Today it stands above 350 ppm. We have increased the concentration, as of 1989, by almost exactly twenty-five per cent. But where did the extra carbon dioxide come from, and when was it emitted into the atmosphere?

Until the late 1980s, these were contentious issues. Researchers could estimate how much fossil fuel had been burnt since the mid-nineteenth century fairly accurately, and they had some idea of how efficient the oceans are at absorbing carbon dioxide. But calculations of how much carbon dioxide might be being released from the biosphere today, and had been released as a result of human activities in the past, were (and are) far less certain. Some researchers suggested that the presence of more carbon dioxide in the atmosphere might be stimulating the growth of existing trees and other plants so much that the biomass was actually taking carbon dioxide out of the air (but they had no way to calculate how big this effect might be); others thought that destruction of tropical forests and other human activities might be putting at least half as much carbon dioxide into the air as burning fossil fuel, exacerbating the greenhouse effect. Meanwhile, oceanographers and chemists

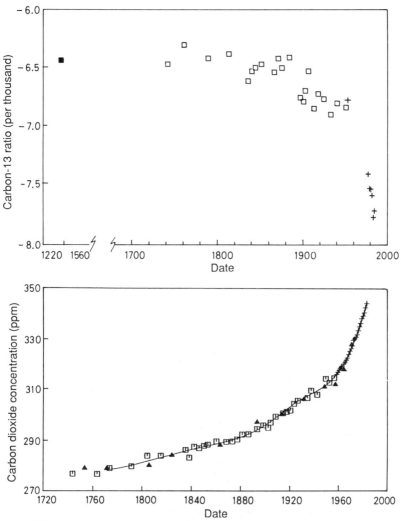

Figure 5.2
Top: *Changes in the proportion of carbon-13 in atmospheric carbon dioxide compared with air from bubbles trapped in ice cores show the effect of the addition of fossil carbon in the twentieth century.*
Bottom: *The ice core bubbles also record the build-up of carbon dioxide going back to the eighteenth century (triangles) and matching up with the Mauna Loa data (crosses).*
(Adapted from figures presented by H. Oeschger and U. Siegenthaler to the Dahlem workshop on 'The Changing Atmosphere', Berlin, 1987. See also Nature, *volume 324 page 237, 1986)*

calculated how much carbon dioxide could be absorbed each year by the seas, and found that this was not enough to account for the fact that only half of the carbon dioxide being produced from combustion of fossil fuels stays in the air. In the late 1970s, this led to a widespread suspicion that there must be a 'missing sink', some reservoir that was absorbing large quantities of carbon dioxide and which had not been included in the calculations. Were the oceans more efficient at doing the job than was thought? Were forests absorbing more carbon dioxide through increased photosynthesis and growth in surviving trees than they were losing as other trees were cut down? Nobody knew. Happily, though, we no longer have to worry about the problem. The Bern researchers have looked at the build-up of carbon dioxide revealed by the changing proportions of carbon isotopes in the air bubbles from the Siple core, and have used this and the computer simulations of carbon cycles to disentangle (or, as they put it 'deconvolve') the separate influences from destruction of the biomass and burning of fossil fuel. In the real world, some plants are indeed likely to be growing more efficiently because more carbon dioxide is available in the air. This carbon dioxide fertilization effect does act to reduce the increase in the greenhouse effect. At the same time, tropical forests are being rapidly diminished in some parts of the world. Some of the trees are burned directly, producing carbon dioxide; some of the once living wood becomes charcoal, storing the carbon before it oxidizes; and the humus and organic litter of the forest floor decomposes more quickly as it is exposed to the air, again releasing carbon dioxide. The beauty of the air bubble study is that it doesn't involve calculation of any of these complexities. All that matters is the net effect – the difference, each year, between the amount of carbon dioxide being released from the biomass and the amount being taken up by other plants; and that shows up directly in the air bubbles.

The carbon cycle models simulate the interchange of carbon dioxide between the oceans and the atmosphere. The researchers used two main models, one with a uniform surface mixed layer, seventy-five metres deep, extending across the simulated ocean, and the other with an 'outcrop' of cold, deep water at high latitudes. The way in which the models allow the oceans to absorb carbon dioxide is determined from measurements in the real world, and especially takes note of the way in which carbon-14 from nuclear weapons tests has penetrated the ocean. Such studies show that if there is a change in the concentration of carbon dioxide in the atmosphere then the equivalent fractional change in the amount of dissolved inorganic carbon at the ocean's surface is only one-tenth as big – there is a 'buffer factor' which limits the amount of extra carbon dioxide the oceans can absorb, and increasing the size of the atmospheric reservoir of carbon dioxide by, say, ten per cent of its present value will increase the size of the reservoir in the surface layer by only one per cent of its present value.

In order to compare the models with the real world, the concentrations of carbon dioxide corresponding to the present situation can be fed into the computer, and the models are then run backwards from the 1980s. Allowance is made for the amount of carbon dioxide put into the air by burning of fossil fuels, and for the rate at which part of this carbon dioxide excess was absorbed by the oceans. This produces a 'prediction' of how the carbon dioxide concentration in the atmosphere should have changed; it almost exactly matches the measurements of air samples from bubbles in the ice, but only back to about 1900. In the nineteenth century, very little fossil fuel was being burnt, compared with the amount burnt in the twentieth century, and these calculations show that if such combustion had been the only source of anthropogenic carbon dioxide then the 'natural' concentration of carbon dioxide in the atmosphere must have been about 295 ppm, and it should have stayed at that level throughout the nineteenth century. The fact that the concentration actually rose from about 280 ppm to 295 ppm between 1750 and 1900 shows that carbon dioxide was being added to the atmosphere from non-fossil sources (*Figure 5.3*). Since the fossil fuel contribution can be determined fairly easily, the non-fossil (biospheric) contribution can be found simply by subtracting the contribution of fossil fuel carbon dioxide from the total. The models suggest that the amount of excess carbon being released as carbon dioxide into the atmosphere each year during the nineteenth century was between 0.5 and 0.8 billion tonnes.

The two main models differ slightly in their interpretation of changes in the twentieth century, because the model with the outcrop at high latitudes is more efficient at sucking away some of the excess carbon dioxide. If a larger fraction is being sucked away, then in order to produce the measured build-up of carbon dioxide more must have been produced; and since the fossil fuel figures are known, the extra must have come from the biomass. Oeschger and Siegenthaler believe that their two models probably indicate upper and lower limits on the actual production of carbon dioxide from the biosphere in the present century, and

111

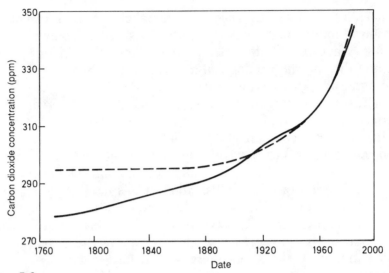

Figure 5.3
By 'backcasting' a calculation of the amount of carbon dioxide put into the atmosphere by burning fossil fuel (dashed line) and comparing this with the measured build-up of carbon dioxide (solid line) it is possible to infer how much carbon dioxide has been released from forest clearance.
(Source: Oeschger and Siegenthaler)

that the truth lies somewhere in between the two sets of figures. In the version with a uniform surface layer, the contribution from the biomass falls to nothing by about 1940; according to the other model, the biosphere has been a fairly steady source of about a billion tonnes of carbon a year. On either calculation, fossil fuel only took over as the main contributor to the build-up of carbon dioxide sometime in the first few decades of the twentieth century; and when the computer models are used to calculate the trend of carbon-13 variations that such changes should have produced, it matches up very closely with the measurements from the Siple core. Up until 1980, the total accumulation of carbon dioxide in the atmosphere from human destruction of biomass amounted to about half of the total produced by burning fossil fuels – but at least fifty per cent of this non-fossil carbon dioxide was added to the atmosphere before 1900. Today, although the contribution from the biomass is between zero and one billion tonnes of carbon each year, the fossil fuel contribution is about 5.3 billion tonnes, and rising.

This study of carbon cycles and reservoirs completely knocks on the head suggestions, taken very seriously in the late 1970s, that destruction of tropical forests was releasing almost as much carbon dioxide into the atmosphere each year as the combustion of fossil fuel. Another line of evidence points in the same direction. If the fraction of excess carbon dioxide staying in the air today is roughly constant from year to year, then the rate at which carbon dioxide is building up in the air can be matched almost perfectly against the rate at which carbon dioxide is being released from fossil fuel. This sounds obvious, but in fact the match is rather delicate, since it involves fuel use that increases from year to year – the graphs involved are curves, not straight lines. The match holds up over the entire interval of the Mauna Loa record, since 1958. But if there were another source of carbon dioxide as well, there should be no simple match between the two sets of figures – the Mauna Loa curve would be distorted in a tell-tale way. Of course, there is some uncertainty in the numbers, but they set a definite limit on the amount of non-fossil carbon dioxide being added to the atmosphere each year. This limit is two billion tonnes of carbon, comfortably above the figures that come out of the Bern study.*

Other researchers have estimated, on the basis of the rate at which forests are being destroyed and new land is coming under the plough, that the rate of release of carbon from the biosphere is about 1.6 billion tonnes a year (perhaps less), and that the total release between 1860 and 1984 was about 150 billion tonnes (but might have been as little as a hundred billion tonnes or as much as two hundred billion). These figures take no account of the carbon dioxide fertilization effect, which will tend to reduce them, though nobody can say by how much. They are therefore very much in line with the figures that come out of the computer models. There is no evidence of any missing sinks at work, and it seems that all the important carbon cycles have already been included in the calculations. This is very important in assessing the value of forecasts of how much carbon dioxide will stay in the air in future decades.

The proportion of carbon dioxide that stays in the air depends, among other things, on how quickly it is released, since the oceans

---

*There is one possible, but implausible, loophole in this argument. More carbon dioxide could be released from non-fossil sources and still not distort the figures, but only if the rate at which this source is growing exactly matches the rate at which fossil fuel use is growing. It seems highly unlikely that the clearance of jungle in Brazil and elsewhere does indeed increase, from year to year over a period of three decades, at precisely the rate at which the use of coal in industry and power stations around the world is increasing.

are slow to respond to a change. The amount of extra carbon dioxide in the atmosphere today produced by burning fossil fuels, mostly since 1940, indicates an 'airborne fraction' for this contribution of about sixty per cent; but the equivalent figure for the *total* input of extra carbon dioxide up to 1980 is only fifty per cent, because the oceans have had time to respond to the slow build-up that was going on before 1940 (some calculations suggest even a slightly lower figure for the airborne fraction). The more slowly we put carbon dioxide into the air, the smaller the fraction of it that will stay there to contribute to the greenhouse effect. But before we look at just how rapid that build-up of carbon dioxide from fossil fuel use is likely to be in the years ahead, I should round off the story of the contribution from forest clearance. It may be small today compared with the contribution from burning fossil fuel, but it is far from being negligible.

Out of the wood

Forests and other living materials are not permanent, or even long term, stores of carbon, but are involved in cycling carbon through the biosphere, like the oceanic and atmospheric reservoirs. The average lifetime of a tree is at least forty years, but the average 'residence time' of a carbon atom in a forest is less than half this, between sixteen and twenty years, because less than half of the carbon fixed by photosynthesis goes into the manufacture of cellulose. The average residence time for carbon atoms in other plants, outside the forests, is about three years. Before human intervention on a large scale, the sizes of forests must have varied, changing the amount of carbon held in temporary storage. But although the individual atoms of carbon being held in storage were changing, and the extent of the forest may have varied, there was always a large pool of carbon that was not immediately available to add to the carbon dioxide concentration of the atmosphere. The size of this forest pool has been drastically reduced over the past two centuries, and is now probably below two hundred billion tonnes of carbon.

Undisturbed soil holds even more carbon than there is in the form of trees. Although ecologists argue about the techniques for measuring carbon in the soil, and the way in which these measurements should be extrapolated to give estimates for the world at large, their various estimates all give figures in the same range,

114

close to 1500 billion tonnes of carbon. Adding in the carbon from forests and from plants other than trees, the total amount available in the biomass, which could in principle easily be converted into carbon dioxide, is about three times as much as the total amount of carbon in the form of carbon dioxide in the atmosphere today (more than 2100 billion tonnes, as against 725 billion tonnes).

The image we usually have of the contribution of the biomass to the build-up of carbon dioxide is of tropical forests in countries such as Brazil being hacked down and burnt. But it isn't just the trees that matter, and it has not always been tropical forests that suffered. When forests are replaced by agricultural activities, at least ninety per cent of the carbon in the form of trees and other plants does indeed get into the atmosphere as carbon dioxide. But this is only part of the story. The natural soil at the base of those forests usually contains as much as forty or fifty per cent of carbon by weight (dry weight – after water has been removed). But soil samples from material turned over by the plough on land that has been used in agriculture for decades contains only a few per cent of carbon as dry weight. Even allowing for the fact that carbon deeper below the surface may not have been disturbed so much, up to half of the original carbon content of the soil must oxidize slowly into carbon dioxide and escape into the atmosphere as new land comes under the plough. This must have been a major contribution to the build-up of carbon dioxide in the atmosphere in the nineteenth and early twentieth centuries. An explosion of pioneer agricultural activity and forest clearance at temperate latitudes took place in the second half of the nineteenth century as new lands in North America, New Zealand, Australia, South Africa and Eastern Europe were opened up by railways, with deforestation and the release of up to fifty per cent of the carbon content of the soil.

One of the biggest single reservoirs of carbon on land today is in the peat of the tundra and high latitude forests. Peat is a form of partially decomposed plant material, saturated with water, that accumulates in swampy regions at temperate and cold latitudes; it is a step in the process by which the carbon of living plants is converted into the coal of fossil fuel, and it is estimated that there is about five hundred billion tonnes of peat around today, most of which has built up at a steady rate since the end of the latest ice age ten thousand years or so ago. This is a useful indication of the relative pace of natural processes compared with human interference in the carbon cycles. It takes a hundred years to accumulate five billion tonnes of carbon in the form of peat; present consumption of fossil fuel releases slightly more carbon than this into the atmosphere as carbon dioxide *each year*.

With so many different elements of the ecosystem to consider, and great uncertainty about exactly how human activities were tilting the natural balance in the nineteenth century, it is no wonder that even the experts have difficulty calculating exactly how much carbon dioxide has been released to the atmosphere from the biomass over the past hundred years or more. Some of the latest calculations suggest that about 150 billion tonnes of carbon was released between 1860 and 1980 (roughly the same as the amount left in forests today), and that the release in 1980 itself was between 0.8 and 1.6 billion tonnes – the figures which, as I mentioned above, seem to agree reasonably well with the latest computer models of how the carbon cycles have changed over that time, calibrated against the ice core isotopes. From 1860 to 1984, the total release of carbon from combustion of fossil fuel was between 168 and 198 billion tonnes (most of it since 1940), and the annual rate of release in the mid-1980s was 5.3 billion tonnes per year, and increasing from year to year. So destruction of the biomass today represents about twenty per cent of the carbon dioxide problem. That is why concern now focuses primarily on the increasing use of fossil fuels around the world, and why projections of the growth of the carbon dioxide greenhouse effect are largely related to projections of the growth in global use of energy from fossil fuels. The best projections do include an allowance for the biomass. As we shall see in the next chapter, those projections are disturbing enough in themselves; but as we shall also see, the carbon dioxide greenhouse effect may itself now be no more than half the greenhouse problem.

SIX

Too Much of a Good Thing

Maybe, as we saw in Chapter Four, a slightly warmer world would be a better place to live in, compared with the little ice age conditions that prevailed in recent centuries. But all the signs are that the world will get a *lot* warmer, very quickly, as we move into the twenty-first century. Already, contributions to the anthropogenic greenhouse effect from other gases are adding as much to the present global warming as the carbon dioxide produced by human activities. Within thirty years, the carbon dioxide greenhouse effect will be much less than half the problem. It was this realization that there is much more to the greenhouse effect than carbon dioxide alone that brought an explosion of research into the greenhouse effect in the middle and late 1980s, and changed all the forecasts of how warm the world is likely to become in our lifetimes.

Although accurate, continuous monitoring of the build-up of carbon dioxide in the atmosphere only began in 1958, it isn't quite true to suggest that there was no interest in the greenhouse effect between the time when Arrhenius and Chamberlain looked at the problem and the time when Roger Revelle and Hans Suess came up with their memorable phrase about a 'global geophysical experiment'. In the late 1930s and early 1940s, following the dust bowl years of the great drought in North America, there was some concern, especially in the United States, at the possibility that the world might be getting warmer. One researcher, G. S. Callendar, published a series of papers at that time in which he looked at the figures for emissions of carbon dioxide from fossil fuels between 1880 and 1935, calculated the greenhouse effect of the presence of this additional gas in the Earth's atmosphere, and compared his calculations with measurements of temperature changes over land and sea during the same period. He suggested that during those decades the increasing concentration of carbon dioxide in the atmosphere should have been warming the world by about 0.03°C per decade, and he compared this with an actual warming of about 0.05°C per decade inferred from the temperature records. Cautiously, Callendar did not claim that the rough agreement between the two sets of numbers proved that the greenhouse effect was at work. He pointed out that 'the course of world temperatures during the next twenty years should afford valuable evidence as to the accuracy of the calculated effect'. In fact, the course of world temperatures over the twenty years from 1940 to 1960 was downward, even though carbon dioxide continued to build up in the atmosphere at an increasing rate. That is why there was little interest in the greenhouse effect until the 1960s and 1970s, when first the Mauna Loa and South Pole measurements began to show just how much carbon dioxide was staying in the atmosphere, and then the course of world temperatures turned clearly upward once again.

There is no doubt that other factors are also at work in the weather machine, nudging the pattern of temperatures up or down from decade to decade. But these factors are now beginning to be understood, and with the help of calculations like those of Wallace Broecker, based on the ice core record of temperatures, we can see why temperatures fell between 1940 and 1960 in spite of the growing greenhouse effect, and how the greenhouse effect in the 1990s and beyond is likely to be so strong that it will overwhelm any natural fluctuations of this kind. In the 1940s and 1950s, a natural downward fluctuation in temperatures was more than enough to overwhelm the embryonic greenhouse effect; in the 1990s, and especially in the twenty-first century, the greenhouse effect will be so strong that no natural downward fluctuation in temperatures will do more than slow, temporarily, the rising trend. Putting that trend in reverse, and cooling the world even a little over the next twenty years, would require a bigger natural fluctuation than anything that has happened since the most recent ice age. Within fifty years, the downward fluctuation in temperatures required even to return the world to the conditions of the 1970s will be bigger than the shift from an interglacial into an ice age – unless policy makers worldwide take immediate action to control the emissions of greenhouse gases.

In our lifetimes, the most important of those gases will always be carbon dioxide, the archetypal greenhouse gas. Because of this, and because it was the first anthropogenic greenhouse gas to be identified, it has received the most attention, and forecasts of the strength of the imminent global warming are often made in terms of the 'carbon dioxide equivalent', even if the warming is likely to be caused by other gases as well. If carbon dioxide is causing half the problem, then the 'carbon dioxide equivalent' effect on temperatures is assumed, in simple calculations, to be the same effect that would be produced by twice as much carbon dioxide as we are actually releasing. In fact, as we shall see, this is an oversimplification. But it still makes sense to start any roll call of greenhouse gases with a detailed look at carbon dioxide itself, and to use this to show how the forecasts of temperature trends are made.

Future carbon dioxide trends

The only way to predict how much carbon dioxide will be released into the atmosphere in the decades ahead is to predict how much energy will be consumed worldwide each year, and what proportion of that energy will be obtained from coal, oil and natural gas. This is, of course, impossible. Forecasting energy demand requires an accurate forecast of economic growth, and an understanding of the links between economic growth and primary energy demand. Nobody can make such forecasts with confidence even for the next ten years, let alone the next fifty. We do not know whether new technological developments will change the demand for energy, whether wars in the Middle East will affect the price of oil, if discoveries of huge new coal reserves in Africa will make energy cheaper than ever, or even whether the growing awareness of the importance of the greenhouse effect will itself encourage governments to restrict the use of fossil fuel. Looking back only two decades, to 1970, it is clear that nobody using data from the 1950s and 1960s could have predicted how energy use would change in the 1970s. Two oil price jumps, in 1973 and 1979, changed the pattern of energy consumption and reduced the growth in consumption of fossil fuels as economies adapted to changing energy costs. And yet, for completely different reasons, the development of commercial nuclear power, which a 1970 forecaster might have expected to benefit from any oil price increase, also slowed down at this time.

There is no way in which 'forecasts' of energy use and carbon dioxide emissions in the twenty-first century can be regarded as accurate, in the way that the flight of an artillery shell or the swing of a pendulum can be forecast accurately in advance from the known laws of physics. Most of the people who are forced to try to assess changes in carbon dioxide concentrations, in order to make climatic predictions prefer to use the word 'scenario' to describe their projections. A scenario is not a forecast of how the world *will* be in twenty, thirty or fifty years from now; it is an image of how things *might* develop, if certain choices are made today. The most useful scenarios are very often, almost by definition, not self-fulfilling but self-defeating. If a scenario shows that continuing unrestricted use of fossil fuel will cause such a strong greenhouse effect that the climatic consequences are perceived as intolerable, then it is almost certain that steps will be taken to reduce carbon dioxide emissions, and the real world will never end up in the state 'forecast' by the scenario. That doesn't prove the scenario was wrong, or make it any less useful; all of the forecasts, predictions and projections discussed in this book are really scenarios, in this sense of the word, whatever term they may be referred to by.

Although nobody can predict how much fossil fuel will be burnt between now and say, the middle of the next century, it is possible to set some limits on the possibilities. The world's population is still increasing rapidly, although the rate at which it is increasing now seems to be slowing down. United Nations' forecasts, which most experts accept as a reliable guide, indicate that there will be six billion people on Earth by the end of the present century, and ten billion by 2100. At the same time, the less developed countries of the world are seeking to emulate the perceived success of the 'rich north' in becoming industrialized and having a comfortable standard of living. This takes energy – power consumed in factories (and to build the factories) to produce the essentials and the luxuries of a better way of life, fuel consumed in transporting goods around the world, energy intensive agriculture to increase crop yields and provide more (and more interesting) food, and so on and so on. It is conceivable, but highly unlikely, that rich countries could use energy more efficiently and at the same time reduce their consumption of energy per person (with a reduction in living standards) while the poorer countries caught up until the whole world was at the level enjoyed by, say, France or Britain in the 1960s. In that case, if steps down the appropriate road were taken at once, the world as a whole might be consuming no more fuel in 2050 than it does in 1990. But I doubt if I need to go into any

120

more details to convince most readers that this possibility represents the extreme limit of any plausible 'low energy use scenario'.

Setting the upper limit on plausible energy scenarios is much more difficult. The best, and most comprehensive, discussion of the possibilities in the context of the greenhouse debate can be found in a weighty study carried out for the United Nations Environment Programme, the World Meteorological Organization and the International Council of Scientific Unions, and published by the Scientific Committee on Problems of the Environment as the twenty-ninth in a series of studies of man-made environmental changes. For obvious reasons, the report is known as SCOPE 29.* It points out that in the early 1980s about seventy-five per cent of the population of the world lived in developing countries, but earned only twenty per cent of the global income. Such countries were then fifty to a hundred years 'behind' the industrialized countries in terms of their energy use today, but may be catching up rapidly in many cases. But what will that catching up involve? In seventy years' time, says the SCOPE report, Brazil may have developed an economy like that of the United States today, but Ethiopia is more likely to resemble present day Spain (while Spain itself, of course, is rapidly being transformed and using more energy per person than ever before). Taking into account all of the uncertainties, and using data from more than a dozen separate scenarios prepared by other researchers, the SCOPE team concluded that growth in the use of fossil fuel at a rate of 2.3 per cent per year until 2050 represents an upper limit – a scenario that is probably as implausible as the prospect of zero growth in global energy demand, but suggests that twenty billion tonnes of carbon, as carbon dioxide, will be added to the atmosphere each year in the 2050s, compared with 5.3 billion tonnes each year today. So how does that compare with past trends?

The amount of carbon dioxide emitted worldwide from fossil fuel combustion grew at a steady rate of just under 4.5 per cent a year from 1950 until the first oil crisis of the 1970s. Between 1860 and 1910, the equivalent growth rate had been about 4.2 per cent a year, but growth was slightly slower between 1920 and 1950. Between 1950 and 1980, the proportion of these emissions from Western industrialized countries fell from sixty-eight per cent

*The full title is *The Greenhouse Effect, Climatic Change and Ecosystems*; it was edited by Bert Bolin, Bo Döös, Jill Jäger and Richard Warrick. Details are given in the bibliography, under 'Bolin' – but I shall continue to refer to it as 'SCOPE 29' for brevity.

to forty-three per cent, while the share from the centrally planned economies increased from nineteen per cent to thirty-three per cent, and the share from developing countries rose from six per cent to twelve per cent (not just a bigger share, but a bigger share of a bigger cake). The figures highlight the way in which industrialized nations have become more efficient users of energy as prices rose in the 1970s, and this trend continued during the economic recession of the early 1980s. But the developing countries have continued to increase their use of energy in spite of everything. Between 1960 and 1973, growth in energy consumption was 7.8 per cent per year in the developing countries, compared with 4.8 per cent a year in the industrialized world; since then, the increase in the industrialized countries has been only 0.5 per cent a year, but use in the developing world is still growing at six per cent a year. Overall, the present rate at which carbon dioxide from fossil fuel is being put into the atmosphere is increasing at 1.5 per cent a year.

Clearly, it is *possible* that growth in energy consumption could return to the high levels of the 1960s and early 1970s, as long as there is enough fossil fuel around in sufficiently accessible locations. The reasons why such scenarios are now considered unlikely have as much to do with changing attitudes as with the physical constraints – one of the influences that seems likely to keep the growth in emissions below about half of the rates recorded between 1950 and 1970 is the awareness of the greenhouse problem itself.

Although accessible reserves of oil and gas are unlikely to last beyond the middle of the next century (unless their use as fuels is curtailed), there is so much coal already located in recoverable deposits that it could fuel energy growth at five per cent every year until the end of the twenty-first century. Altogether, *known* exploitable reserves contain at last ten times the amount of carbon present in the atmosphere as carbon dioxide today. And the regions in which these reserves lie represent only a small part of the Earth's crust. Very little exploration for coal has been carried out in many parts of the world, such as Africa and the heartlands of the Soviet Union, simply because there has been no need to look for more.

Against this background, it seems likely that emissions of carbon dioxide will continue to increase in the immediate future. The highest possible rates of increase are unlikely to happen, because some countries, such as the United States and the members of the European Community, are beginning to perceive the greenhouse effect as a threat, and will adjust their policies to reduce emissions (or, at least, hold them steady). At the other extreme, a country such as China, with enormous reserves of coal, a large industrialization programme, a quarter of the world's human population and a commitment to improving living standards, is likely to press ahead regardless. Some countries, indeed, may perceive climate changes caused by the greenhouse effect as beneficial to themselves, and therefore do nothing to limit carbon dioxide emissions whatever the detrimental effects elsewhere.

To achieve either the high growth scenario or the low growth scenario would require international cooperation, with all countries working for the same end. But with some pushing one way and some pushing the other, the cynics argue that the simplest assumption is that overall growth in carbon dioxide emissions will continue at roughly the present rate. Another simple way of looking at the size of the carbon dioxide problem is to think in terms of total emissions over the next few decades, rather than the rate at which emissions might increase from year to year. If half the available reserves of fossil fuel known today were burnt over the next hundred years, the concentration of carbon dioxide in the atmosphere would rise above a thousand ppm. If, on the other hand, consumption of fossil fuel simply doubles over the next hundred years, that atmospheric concentration will still exceed six hundred ppm. The figures are striking, but they are no substitute for a guide, no matter how approximate, to how the concentration of carbon dioxide may increase year by year in our lifetimes. *Figure 6.1* shows a range of plausible scenarios for the amount of carbon dioxide *staying* in the air, suggested by an Australian climatologist, Graeme Pearman. He takes the SCOPE 29 upper limit and compares this with scenarios for continued growth in emissions at the present rate and, as a low scenario, continued growth at the present rate until 2000, followed by steady release of carbon dioxide at the same rate each year.* The interesting comparison is with a concentration of 560 ppm, twice the 'pre-industrial' concentration of carbon dioxide in the atmosphere. But his calculations of the amount of carbon dioxide

---

*It is important not to confuse a steady rate of emissions (the same amount each year) with a steady *growth* in the rate of emissions (the same fractional increase each year). When the rate of inflation falls from seven per cent a year to six per cent a year, that doesn't mean that the goods in the shops are now one per cent cheaper – all it means is that prices have gone up by six per cent this year, *on top of* the seven per cent they went up last year. Growing emission rates are like inflation – they build on the growth from previous years.

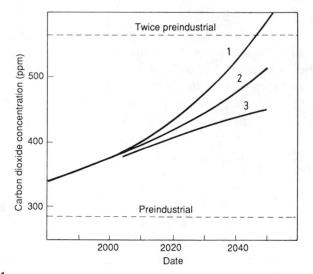

Figure 6.1
Graeme Pearman's calculation of the way the carbon dioxide concentration of the atmosphere will increase (1) if growth continues at 2.3 per cent a year, (2) if growth continues at 1.5 per cent a year, and (3) if growth continues at 1.5 per cent a year until 2000, and releases then continue at the same rate each year. (Adapted from Greenhouse, *edited by G. Pearman)*

in the atmosphere in the twenty-first century depend not only on scenarios of how much is likely to be released, but also on calculations of how much is being absorbed by the oceans.

The balance between the amount of carbon dioxide produced and the amount that stays in the air is worked out using models like those of Oeschger and Siegenthaler, mentioned in the previous chapter. They showed that if emissions of carbon dioxide were held steady at the 1975 level, equivalent to an additional 2.4 ppm of carbon dioxide put into the air each year, the fraction of the additional carbon dioxide staying in the air would be between thirty-eight and fifty-nine per cent in 2070, and a doubling of the pre-industrial concentration would not be reached before 2160. In their work, Oeschger and Siegenthaler picked a far more extreme upper limit than the SCOPE researchers, and calculated the effects of all-out burning of coal at a growth rate of five per cent a year. In that scenario, the carbon dioxide concentration doubles by the year 2020, and soars to five or ten times the pre-industrial concentration at the end of the twenty-first century. For any intermediate case, between forty-eight and eighty-three per cent of the total carbon dioxide released would still be in the air a hundred years from now, and as the concentration increases the oceans become less effective as a carbon dioxide sink.

For their middle course, though, Oeschger and Siegenthaler do not make any assumptions about how much fossil fuel might actually be burned. Instead, they work backwards to set a limit on how much we could allow ourselves to burn, if we want to keep the build-up of carbon dioxide within reasonable limits. It all depends, of course, on what you regard as reasonable. Purely for the sake of argument, they set their limit at fifty per cent over (1.5 times) the pre-industrial concentration – 420 ppm. That constraint could be met even if *growth* in emissions continued for ten more years, peaking at a release of the equivalent of three ppm in 2000. But then the use of fossil fuel would have to fall dramatically, back to 1970 levels by about 2030 and levelling off at about twenty-five per cent of the 1970 rate at the end of the twenty-first century. 'Around the turn of the [present] century,' say the Bern researchers, 'new technologies would have to take over.' If you think that is going to happen in the real world, you are certainly more optimistic than I am. They also make a very telling point which is often ignored in simpler forecasts.

The buffer factor of sea water increases as the concentration of carbon dioxide in the atmosphere increases, so the oceans become *less* effective as a sink for the gas. If all fossil fuel reserves were exploited and burnt, as in their high scenario, as much as eighty per cent of the carbon dioxide would stay in the atmosphere in the short term. Once the release of more carbon dioxide stopped, the concentration would very slowly fall until it reached a new equilibrium state. But that would *not* be the same as the equilibrium that existed from the end of the latest ice age until the beginning of the nineteenth century. At equilibrium, the atmosphere would still contain at least one-eighth of the carbon dioxide released as a result of human activities. The same effect, on a smaller scale, applies for the less dramatic scenarios. If we wait until we can see the harmful impacts of the carbon dioxide greenhouse effect on climate, and then halt further emissions immediately, it will still be too late – the atmosphere will *not* then return to its pre-industrial state, no matter how long we wait.

This is why it is so important that climatologists try to model future climatic trends on the basis of the information available now. The scenarios may be only a rough guide to how carbon dioxide concentrations will increase, but they are all the guidance we have, and if the changes in carbon dioxide concentrations are likely to cause unwelcome climatic changes then we have to take action now, not after those changes have begun to occur. The energy use scenarios, for all their imperfections, tell us that the carbon dioxide concentration is likely to double (compared with the 'natural' concentration) sometime in the next fifty to seventy-five years, and is almost certain to increase to fifty per cent above the pre-industrial level within thirty years. And the computer modellers can use that information to give us a good guide to how the world will warm as a result.

Making the most of the models

Computer simulations of climate – climate models – use the basic equations of physics, chemistry, and even of biological processes to calculate how the weather systems of the world work. None of this could be done until reasonably powerful electronic computers became available in the 1960s. Almost all of the present day models are descended, conceptually, from the pioneering work of Syukoro Manabe (sometimes referred to as the 'godfather' of climate modelling) and Richard Wetherald at the Geophysical Fluid Dynamics Laboratory in Washington, DC. They developed the idea of a global energy balance between the amount of heat reaching the Earth from the Sun, the amount being radiated from the ground, the amount being trapped in the atmosphere and the amount finally escaping into space. In equilibrium, the amount arriving from the Sun and the amount escaping from the top of the atmosphere are in balance, but atmospheric feedbacks make the surface of the Earth warmer than it would be if the planet had no atmosphere. In their earliest study, published in 1967, Manabe and Wetherald arrived at an estimate of a rise in global mean temperatures of 2°C for a doubling of the present day concentration of carbon dioxide. Although more sophisticated models that use bigger and faster computers and include more feedback effects now suggest that the actual rise in average temperatures for a carbon dioxide doubling might be twice that, modern models that concentrate on the same basic features of the carbon dioxide greenhouse effect that Manabe and Wetherald studied in the 1960s still give similar forecasts.

The models used today actually vary enormously in the amount of detail they provide, depending on the computer that is being used to run them and the feature of the climate that is being studied. The simplest models, for example, can calculate the average temperature at the surface of the Earth from the equations that describe the balance between the Earth's reflectivity, the amount of energy coming in from the Sun, and the average radiation properties of the Earth's atmosphere – which wavelengths of energy it absorbs, and which ones it allows to pass. Such a model is called 'zero-dimensional', because the real temperature pattern over the surface of the globe is replaced by a single number, as if the globe were shrunk to a point. Zero-dimensional models are simple, but useful. They explain, for example, why the surface of the Earth is warmer than the surface of the Moon, and show the natural greenhouse effect at work.

At the other extreme of the modeller's craft, a three-dimensional model contains a set of numbers corresponding to the temperature at different points, not only over the surface of the Earth (at different latitudes and longitudes), but also at different altitudes through the atmosphere. The most sophisticated three-dimensional models are known as general circulation models, or GCMs. If they are fed with a set of numbers defining the temperature and other physical properties of the atmosphere over a three-dimensional grid covering the whole globe, they will calculate how the pattern of temperature, humidity, wind speed and direction, sea ice, soil moisture and other climatic variables will change as time passes. All this requires a very powerful computer, and a lot of computing time. One typical GCM, run by Stephen Schneider and his colleagues at the US National Center for Atmospheric Research (NCAR), divides the surface of the Earth up into a grid of 1,920 rectangles, a lattice in which each rectangle covers 4.5° of latitude and 7.5° of longitude (at a latitude of 40° – near Madrid, Beijing or Indianapolis – this corresponds to a rectangle of five hundred by 640 kilometres). Then the atmosphere above each rectangle is divided into nine layers, giving a total of 17,280 boxes filling the atmosphere to an altitude of thirty kilometres. The climate parameters (temperature and so on) are specified at the corners of each box. Starting from an appropriate set of numbers spread through this grid to describe the instantaneous pattern of weather around the globe, a Cray XMP supercomputer can calculate a year's 'weather' in simulated thirty-minute steps in about ten hours of real time.

Clearly, this is only a very broad guide to weather patterns, even in the simulated world. The GCM cannot say anything in detail about what is happening inside each box, and there are many countries in the world that would fit completely inside a single box.

Important climatic phenomena, such as clouds, are much smaller than the size of each box, and cloud cover has to be represented as an average over each box. This might mean, for example, that in such a simulation of global climate there is either no cloud cover at all over England, or a uniform layer of cloud. The limitation with a GCM is that it cannot tell you much about detailed weather patterns. But the GCMs *do* give a good guide to average conditions over the Earth. The first check on them, of course, is to see how well they reproduce the patterns of wind, rainfall and temperature that occur in the real world. They do it so well, in fact, that they reproduce beautifully the cycle of the seasons, 'predicting' a pattern of temperature differences between August and February for each region of the globe that closely matches the broad features of the real world. This is an even more impressive test of the models than it may seem at first sight (to anyone used to the cycle of the seasons), since the temperature changes involved are, on average, several times bigger than the temperature changes involved in the shift from an ice age to an interglacial and back again – every year, the opposite hemispheres of the Earth each experience a taste of ice age conditions, which we call winter.

Another way to test the GCMs, and to begin to use them to develop an understanding of the Earth's climate, is to see if they can reproduce patterns of past climates. Why, for example, was the world so warm during the climatic optimum, around six thousand years ago? John Kutzbach and colleagues at the University of Wisconsin at Madison used a GCM to calculate temperature patterns on Earth at that time, allowing for the fact that the tilt of the Earth's axis was slightly greater then and that the Earth made its closest approach to the Sun in June, not in January as it does today. Both these orbital effects (fine tuning of the Milankovitch rhythms) increase the contrast between seasons in the northern hemisphere, together making summer five per cent hotter and winter five per cent colder than today. Kutzbach's team found that the equivalent change in their GCM produced just the pattern of increased summer temperatures and more intense monsoon rainfall in the northern hemisphere that is revealed by geological evidence from the time of the optimum.

Schneider himself, working with Starley Thompson at NCAR, carried out a similar simulation of the short-lived return of cold weather around the North Atlantic about eleven thousand years ago, after the ice age glaciers had begun to retreat. They followed up a suggestion, made by several climatologists, that this thousand-year-long cold spell was, ironically, caused by the break-up of the ice sheets. The argument is that large amounts of fresh water, from the melting ice, would have poured into the Atlantic Ocean, forming a lid over the denser, salty water below. Because fresh water freezes more easily than salt water, this lid could have frozen solid, stopping the heat from the warm water current known as the Gulf Stream from escaping to warm the air over northwestern Europe. The NCAR researchers tested the idea by running a computer simulation in which the Atlantic was specified as frozen down to latitude 45°N, and found just the pattern of regional and seasonal changes in temperature and wind patterns that is inferred from the geological record.

All this is fascinating in its own right, and helps us to understand past patterns of climatic change. But the important point for our purposes now is that these successes tell us that the GCMs are working along the right lines. So when the same GCMs are set the task of calculating how the temperature and rainfall patterns of the world will change when there is more carbon dioxide in the air, we can accept that they will be giving us a genuine guide to the way the real world will change. These are the models that 'predict' that the world should have warmed, on average, by about 1°C over the past hundred years as a result of the carbon dioxide greenhouse effect. This agrees fairly well with the actual warming described in Chapter One, once allowance is made for the way the oceans take up some of the extra heat and slow the warming. So GCMs can explain seasonal variations today, broad features of climatic changes that occurred thousands of years ago, and the main feature of changes that have happened over the past hundred years. What can they tell us about the likely climatic changes of the next few decades?

The greenhouse champion

The surface of the Earth receives, on average, about eighteen per cent of its heat directly from the Sun, and a little less from solar radiation scattered on the way through the atmosphere by clouds; some sixty-five per cent of the radiation reaching the ground is in the form of infrared energy radiated back down to the surface by the atmosphere – the natural greenhouse effect. Looking at the amount of energy flowing out from the surface of the Earth, we can see the size of the greenhouse effect in terms of numbers that are reasonably familiar in everyday life. A typical household light bulb may radiate a hundred watts of energy, chiefly in the form of light.* Each square metre of the Earth's surface, averaging over the entire globe and a whole year, radiates 390 watts. But the amount of radiation escaping into space, measured by satellites, amounts to only 237 watts per square metre. This, of course, exactly balances the incoming energy from the Sun (or, rather, the incoming and outgoing energy flows *did* exactly balance, until human activities started changing the composition of the atmosphere), but the difference, 153 watts per square metre, represents the warming of the Earth by the greenhouse effect (*Figure 6.2*). The natural greenhouse effect is roughly equivalent to warming the surface of the Earth with an array of one-kilowatt electric heaters spaced out over the entire surface of the globe so that each fire is warming an area of six and a half square metres – about the size of a carpet ten feet (3.2 metres) long and six feet (two metres) wide. Nitrogen, the main constituent of the atmosphere, and oxygen, the second largest component, play no part in this natural greenhouse effect at all. It is entirely due to the presence of a trace of carbon dioxide, and to a feedback between temperature and the amount of water vapour in the air.†

These feedbacks are included in the best GCMs. Carbon dioxide absorbs outgoing infrared radiation in the waveband from thirteen to seventeen micrometres, but water vapour absorbs across the infrared spectrum, from above twenty-five micrometres to below six micrometres.‡ Even so, before human activities started altering the natural balance, almost eighty per cent of the energy radiated from the ground in the waveband from seven micrometres to thir-

---

*A watt is a measure of the amount of energy flowing out each second; all the figures here refer to continuous energy flows, *not* the total over a year.

†Incidentally, some people worry that a build-up of carbon dioxide may be removing large quantities of oxygen from the air. True, each molecule of carbon dioxide does sequester two atoms of oxygen, leaving that much less around for us to breathe. But more than a fifth of the atmosphere is oxygen, and even increasing the carbon dioxide content of the air tenfold, from 0.035 per cent to 0.35 per cent, would be like taking no more than a drop of water out of a reservoir.

‡Water *vapour* is not the same thing as clouds. The vapour is a transparent gas that traps heat. Clouds are made of water droplets or ice crystals, and reflect heat. Both are taken into account in the best models.

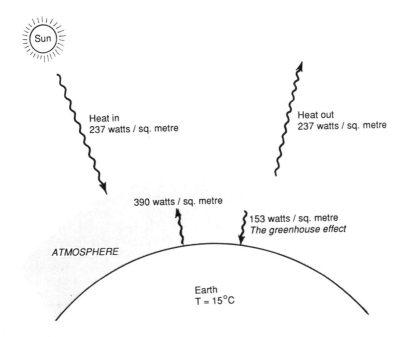

Figure 6.2
The natural greenhouse effect expressed in terms of the amount of heat crossing each square metre of the Earth's surface. The build-up of anthropogenic greenhouse gases in the atmosphere reduces the outward flow of heat from the top of the atmosphere and increases the amount returning to the ground; as the world warms as a result, the outward flow increases once again, and if the build-up of greenhouse gases halted then a new equilibrium, with a higher surface temperature, would become established. But the build-up of greenhouse gases shows no sign of being brought to a halt.

teen micrometres escaped into space – this is the region known as the 'window' in the infrared spectrum (*Figure 6.3*). Many of the gases released by human activities absorb in this region, 'dirtying' the window. Adding carbon dioxide to the air, however, simply strengthens the natural processes that already keep the world warm.

Even doubling the amount of carbon dioxide in the atmosphere, from three hundred to six hundred ppm, only increases the amount of heat trapped by the greenhouse effect by an average of four watts per square metre. The increase is so small because the concentration of carbon dioxide is so large that most of the heat trying to escape from the Earth in the waveband from thirteen to seventeen micrometres is trapped already – radiation that has been trapped by one carbon dioxide molecule cannot be trapped

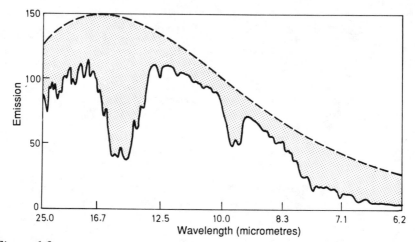

Figure 6.3
The amount of infrared radiation at different wavelengths emerging from the Earth's atmosphere has been recorded by instruments on board satellites such as Nimbus 4. These particular measurements were made over the tropical region of the Pacific Ocean when there were no clouds. The dashed line shows the radiation that would be produced by an object with a temperature of 27°C, about the temperature of the surface in this region. The amount by which the actual emission falls short of this indicates the absorption of infrared in the atmosphere — the greenhouse effect. Water vapour absorbs across this band; carbon dioxide is responsible for the strong absorption between thirteen and seventeen micrometres. There is a 'window' between seven and thirteen micrometres where most of the radiation from the Earth's surface escapes into space.
(Source: NASA data, presented to the Dahlem meeting by V. Ramanathan)

by another one. But even this modest increase in the strength of the greenhouse effect will cause the world to warm slightly, and that will evaporate more water vapour from the oceans. Because it absorbs across the infrared spectrum, and because it is released from the seas in large quantities as the world warms, water vapour is the single most important greenhouse gas. The modest warming caused directly by a carbon dioxide doubling releases water vapour, which warms the world a little more and releases more water vapour. In some computer models, this feedback can be switched off, and the temperature increase without the feedback compared with the equivalent increase calculated with the feedback. Such studies show that the water vapour feedback effect amplifies the temperature increase by a factor of three. In the real world, the feedbacks cannot be switched off. Water vapour makes the carbon dioxide greenhouse effect three times stronger.

The best GCMs include other feedbacks as well. Sea ice,

covered by snow, is a very good reflector, and bounces a lot of incoming solar heat away from high latitudes, especially in winter. Open sea absorbs about twice as much incoming solar energy as the ice does. When the ice melts in the spring, the amount of extra solar radiation being absorbed is at least fifty watts per square metre, more than ten times the local change in heat balance caused by a doubling of carbon dioxide. Some GCMs include the effects of increased melting of sea ice in their calculations, in response to the rise in temperatures predicted for the direct greenhouse effect. This makes a big difference to the temperatures in those models at high latitudes, but only amplifies the global average warming by about ten or twenty per cent, because the effect only operates over a limited area of the globe.

Some models also try to take account of changes in cloud cover. Clouds play a double role in determining the heat balance of the Earth. On the one hand, they reflect away incoming solar radiation – without clouds, our planet would reflect about ten per cent of the incoming solar energy; with clouds, it reflects about thirty per cent. On the other hand, though, some clouds are very good at trapping infrared energy – perhaps contributing as much as fifty watts per square metre, averaged over the globe, to the greenhouse effect. GCMs do not yet deal adequately with clouds, and this is probably their greatest imperfection.

When all the feedbacks are fed in to the GCMs, they predict that the global average temperature increase for a doubling of the amount of carbon dioxide in the atmosphere will be between 3.5 and 4.5°C. Simple models, that ignore some or all of the feedbacks, give a lower figure for the likely global warming, with estimates ranging down to about 1.5°C. Because of this spread of forecasts, many people use a figure of 3°C as a 'best guess' for the temperature increase associated with a carbon dioxide doubling, and the whole range from 1.5 to 4.5°C is often quoted as the range of uncertainty in that estimate. Such figures even appear in many official reports. But this is based on the false assumption that all the forecasts are as good as each other. In fact, the more complete GCMs consistently give estimates at the higher end of this accepted range, and the estimates have got higher as more computer power has become available and more feedbacks have been included. The SCOPE 29 report assessed the strengths and weaknesses of all of the main computer models used in this kind of forecasting, and reached the conclusion that, including feedbacks, the most likely temperature increase for a carbon dioxide doubling, once the world has had time to settle down into a new equilibrium, will be 3.5°C, and that the range of uncertainty extends from 2.5 to 4.5°C.* The GCMs that produce a warming in this range also suggest an increase in precipitation, worldwide, of between seven per cent and eleven per cent, as some of the extra water vapour that gets in to the air turns back into clouds, and falls as rain and snow. But that average increase masks the fact that some regions – near coasts – will get much more rain, while others – in the continental interiors – become much drier than they are today. Before we can begin to forecast when such changes might take place (how soon the world will experience a warming equivalent to the effect produced by a doubling of carbon dioxide) we need to know just how much other gases are contributing to the anthropogenic greenhouse effect. Carbon dioxide is still the champion, but there are strong contenders for the crown. It is time to meet them.

Methane and the main contenders

Apart from carbon dioxide (and the feedback with water vapour), there are three 'natural' greenhouse gases which are building up in the air as a result of human activities, and a family of gases that also trap heat and warm the Earth, but which are produced only by human industrial processes, not by any natural activity. The other natural greenhouse gases are methane, nitrous oxide, and ozone (in the lower part of the atmosphere, the troposphere). The unnatural contribution to global warming comes from compounds called chlorofluorocarbons, or CFCs. These now rank second only to carbon dioxide in their warming effect, but are more familiar as the villains of the story of the 'threat to the ozone layer', in the stratosphere tens of kilometres above our heads. Ozone that is threatened in the stratosphere, and which is essential there for the well-being of Gaia, is the same chemically as the ozone that is both a poison and a greenhouse gas, contributes to acid rain and photochemical smog, and is therefore unwelcome at lower altitudes. The connection between CFCs and ozone is one of the

*In my view, this is still too kind to the simple models. I'd go along with a 'best guess' of 3.5°C for a carbon dioxide doubling, but put the range of uncertainty from 3°C to 5°C, since there may be feedbacks that have not been included yet.

most complicated tangles in the whole greenhouse effect saga; the roles of methane and nitrous oxide are, by comparison, easy to understand.

Methane (CH_4) is the most abundant hydrocarbon in the atmosphere, and is a waste product of the biological activity of some kinds of bacteria. Many of these bacteria live in swamps, where bubbles of methane gas often escape and sometimes burn spontaneously as the flickering blue flames of the will o' the wisp. It is also known as marsh gas. Other bacteria that produce methane live in the guts of animals, especially ruminants, and help to break down the food the animals eat into a digestible form.* The rear ends of cows and other animals, and termite colonies, are also prolific sources of methane. But the presence of methane in the atmosphere in significant quantities was only noticed in the 1940s, when spectroscopic studies showed the presence of the character-istic absorption lines of the methane molecule, its spectroscopic fingerprint. It was not until the 1960s that accurate measurements of the concentration of methane in the air could be made, and since then continuous monitoring of the gas has shown that concentration steadily increasing, from about 1.4 parts per million at the end of the 1960s to 1.7 parts per million at the end of the 1980s. The rate of increase is about seventeen parts per billion a year, equivalent to one per cent of the present concentration.

Like carbon dioxide, methane is trapped in air bubbles in ice cores. Studies of bubbles from the Greenland icecap show that the concentration of methane in the air was the same from the early centuries of the present interglacial, ten thousand years ago, up until about three hundred years ago. This natural concentration of methane in the air was about 0.7 parts per million. Over the past three hundred years, the concentration has increased almost exactly in line with the growth in the human population of the world (*Figure 6.4*) – but don't jump to any conclusions. Although human beings do produce methane, the amount is no more than fifty grams a year per person, just about one-thousandth of the output of a healthy cow. The reason that methane in the atmosphere has increased in step with the growth in human population is that people control most of the sources of methane today, and there

---

*They don't do this for the good of the animals, but because the stomach of a cow, say, is a nice warm, safe place for bacteria, continually supplied with food from the mouth end of the animal. It is a working relationship that benefits both parties, a neat example of evolution at work.

is a close link between methane and agriculture. More cattle, to feed a growing human population, make their contribution to the methane build-up; paddy fields, developed for growing rice, are in effect artificial swamps, where the bacteria that excrete methane spread rapidly and make their homes; and burning of wood as fuel, or in slash and burn agriculture, also contributes its quota.

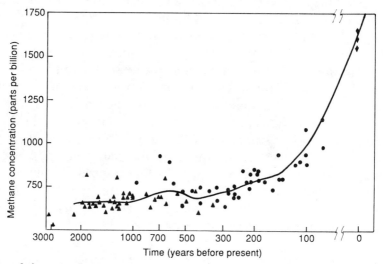

Figure 6.4
The build-up of methane in the atmosphere, determined from measurements of air trapped in bubbles in ice cores.
(Data supplied by Ralph Cicerone)

Methane production is also linked directly with the principal source of anthropogenic carbon dioxide today, extraction of fossil fuel from the ground. Because methane contains carbon, the proportion being released by biological activity can be compared with the amount being released from fossil sources by measuring the proportions of the different carbon isotopes in the gas in the air; such studies suggest that a quarter of the methane being released today does indeed come from fossil sources, carbon compounds that have been locked away below ground for millions of years. This is no real surprise. Methane is a major component of the natural gas associated with oil reserves, and often escapes into the air as an unwanted accompaniment to the extraction of oil. It seeps out of coal seams, where it goes by yet another name, fire damp, and is a potentially explosive hazard to miners. And it is also produced by biological activity (those bacteria again) in municipal and industrial

136

wastes in landfills, a growing source of methane in our throwaway society, which may now be contributing nearly one fifth (almost as much as fossil sources) to the build-up of the gas each year. But the isotope studies show that there is another source of 'old' methane, as well as gas from coal and oil extraction.

Carbon present in the atmosphere in the form of carbon dioxide comes in three varieties. There are two stable isotopes, carbon-12 (the most common) and carbon-13, and one radioactive isotope, carbon-14, that is produced by the action of cosmic rays on atoms of nitrogen-14 in the air. Of the two stable isotopes, the biochemistry of photosynthesis favours the uptake of the lighter atoms by plants, so plant material contains a smaller concentration of carbon-13 than non-living material. When that plant material burns, some of the carbon in it is locked up in molecules of methane. Measurements of the methane trapped in ice bubbles from cores drilled in both Greenland and Antarctica show that the proportion of carbon-13 has decreased in recent decades, and imply that at least 50 million tonnes of methane is now being produced each year by burning biomass, most probably from the destruction of the tropical rainforest.

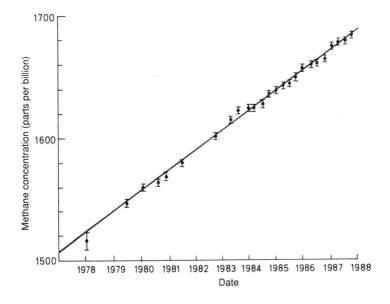

Figure 6.5
A more detailed look at the recent increase in the concentration of methane in the air, measured directly.
(Data supplied by Ralph Cicerone)

Studies of carbon-14 in atmospheric methane give another insight into its origins. Methane from burning biomass, or from the biological activity going on in paddy fields, cow guts and termite mounds, is relatively rich in carbon-14, because it is in equilibrium with the carbon of the atmosphere. But very old sources of carbon, such as coal, contain no carbon-14, since it has all decayed (the half-life of carbon-14 is a little under 6,000 years). Deposits of organic material, such as peat, that are a few hundred or a few thousand years old contain a proportion of carbon-14 that reflects their age, and so does any methane they emit.

Measurements of the amount of carbon-14 in atmospheric methane today show that there must be a large source of old methane getting into the air. This cannot be explained solely by the known emissions from coal mines and oil wells. The conclusion reached by researchers such as Ralph Cicerone, of the US National Center for Atmospheric Research, is that global warming is causing the release of old methane from some natural reservoir.

There are two candidates for this methane source. One is the frozen tundra of the Arctic. As the world warms and the tundra thaws, this peaty reservoir may be releasing methane in increasing quantities. The other reservoir lies beneath the oceans, where methane is locked away in large quantities in the form of compounds known as clathrates, or methane hydrates. These compounds are present in the mud at the bottom of the Arctic Ocean, and in the deeps of ocean trenches. They are kept stable by a combination of low temperatures and high pressures. But if either the pressure above them eases, or the temperature rises, methane is released from the clathrates.

In both cases, there is a positive feedback, reinforcing the greenhouse effect. Global warming releases methane from tundra and from the deep ocean; the build-up of methane makes the world warmer.

Proportionately, this build-up is far greater than the build-up of carbon dioxide in the air. Human activities have increased the concentration of carbon dioxide by twenty-five per cent, compared with its natural level during the present interglacial; but we have more than doubled the amount of methane present in the atmosphere.*

*While the population of the world has quadrupled. Anthropogenic sources of methane have also increased by a factor of four, but about half the methane in the air is of natural origin, so the overall increase produces a doubling of the natural background.

This is a dramatic disturbance of the natural balance of chemical processes involving methane, of which there are many. I don't want to go into all the details here, but some of the subtleties are worth hinting at. Methane reacts strongly with, among other things, oxides of nitrogen and what is known as the hydroxyl radical (OH), which is effectively a water molecule from which one hydrogen atom has been removed. The amount of hydroxyl in the air does not seem to be increasing, so as methane builds up there is proportionately less OH available for it to react with, and a bigger proportion of the methane stays in the air. Some of these reactions also affect the amount of ozone in the atmosphere, both in the troposphere (near the ground) and in the stratosphere. Through reactions like these, methane actually contributes about six per cent of the carbon dioxide added to the atmosphere by human activities each year – *all* the methane eventually gets converted into something else, but the concentration in the atmosphere continues to build up, because it is being released faster than it is being destroyed. Atmospheric methane is a very interesting and important substance. But what we are interested in now is its ability to trap infrared radiation that would otherwise escape into space.

Methane absorbs strongly at 7.66 micrometres, in the 'window' in the infrared spectrum, and the direct heating effect of increasing the concentration of methane from 0.7 ppm to 1.7 ppm is about half of the simultaneous warming caused by increasing the carbon dioxide concentration of the atmosphere from 280 ppm to 350 ppm. Methane may have played a part in the ice age/interglacial cycles of the past, since analyses of air bubbles from the ice cores show that the concentration of methane during an ice age is only about half the concentration during an interglacial. The shift from a concentration of about 0.35 ppm twenty thousand years ago to 0.65 ppm fourteen thousand years ago can easily be explained, since as the world began to warm out of the ice age and glaciers started to retreat there would have been more wetlands around for the bacteria to thrive in, while the rising temperatures may also have stimulated their activity. Their contribution to the warming that began the interglacial was small, about 0.1°C, but by no means insignificant, and provides another reminder of the complexity of the web of interactions which sustains Gaia.

If the build-up of methane in the atmosphere continues to match the increase in world population, then the concentration will reach 2.0 ppm in 2000 (three times the natural concentration for an interglacial), and 2.5 ppm in 2050, with corresponding increases in the strength of methane's contribution to the greenhouse effect. By the middle of the twenty-first century, the change in methane concentration of the atmosphere caused by human activities is likely to be more than six times as large as the change associated with the shift from ice age to interglacial conditions. But there is no point in expressing that in terms of a temperature increase until we have brought into play all the other participants in the warming.

Nitrous oxide and ozone

Nitrous oxide (N_2O; laughing gas) is also building up in the air, although not as rapidly as methane. It is a more minor constituent of the atmosphere, and is measured in parts per billion (thousand million), not parts per million; the concentration in 1976 was 298 ppb in the southern hemisphere and 299 ppb in the northern hemisphere; by 1980 these figures had risen to 301 ppb and 302 ppb respectively. That corresponds to an annual average rate of increase of only 0.2 per cent a year. There are very few accurate measurements of the concentration of nitrous oxide in the air from earlier decades, and it was only discovered in the atmosphere in 1938, but by extrapolating these figures, and some from the 1960s, backwards (and adding in their calculations of how much nitrous oxide is released by human activities) researchers infer that the pre-industrial concentration was about 280 ppb – almost exactly one-thousandth of the pre-industrial concentration of carbon dioxide in the air. The increase is mainly a result of agricultural activity. Plants need nitrogen in order to grow, but most of them are unable to extract it directly from the air, which is almost four-fifths nitrogen. So farmers use nitrogen-based fertilizers (such as nitrates) to boost crop yields, and some of the nitrogen from these compounds gets into the air in the form of nitrous oxide. Oxides of nitrogen, including N_2O, are also produced when anything, including fossil fuel in a car or a power station, is burned, because the heat encourages reactions between nitrogen in the air and oxygen; it is also produced naturally by lightning and by biological activity in soils, and from the oceans.

Although the amount released each year is small, the lifetime of a nitrous oxide molecule in the air is about 170 years, so each one stays around, adding its contribution to the warming of the globe, for a long time. Unlike methane, nitrous oxide does not seem to be involved in any major chemical reactions in the troposphere, but

140

finds its way up into the stratosphere, where it does take part in a series of reactions that ultimately cause a reduction in the concentration of ozone there – a depletion of the ozone concentration of the stratosphere of about four per cent would occur for a doubling of the concentration of nitrous oxide. But that would take a very long time. If the amount of nitrous oxide now being released into the atmosphere each year is maintained, a balance will be struck when the concentration reaches five hundred ppb, and the new equilibrium will be established after about two hundred years. On the more likely assumption that the amount of fossil fuel burnt doubles over the next hundred years, and that nitrate fertilizer use continues to increase, though not as rapidly as during the past two decades, the concentration of nitrous oxide in the atmosphere will reach at least 360 ppb, perhaps four hundred, by 2030. Compared with the pre-industrial concentration, that will represent an increase of about thirty per cent. Since nitrous oxide molecules also absorb infrared radiation in the region of the atmospheric window, this too will make a contribution to the anthropogenic greenhouse effect as we move into the twenty-first century – a contribution perhaps half as big as the contribution being made by methane over the same period of time.

Of all the natural greenhouse gases, however, the hardest to include in any warming scenario is ozone. Ozone is a form of oxygen in which each molecule contains three atoms (O_3) instead of the more common di-atomic form (O_2), the kind that we need to breathe in order to live. It is highly reactive, and is involved in interactions with all the gases I have mentioned so far, as well as the CFCs, which we are coming on to shortly, and others besides. Ozone reactions very often involve sunlight as well – they are photochemical reactions – and ozone is unique among the greenhouse gases in that it not only absorbs infrared radiation going out from the surface of the Earth, but also ultraviolet radiation coming in from the Sun. Most (ninety per cent or more) of the ozone in the Earth's atmosphere is in the stratosphere, where it is both created and destroyed by photochemical processes in a series of reactions that are roughly in balance naturally, but which are now being disturbed by human activities.

Ozone is produced in the troposphere by reactions that involve oxides of nitrogen, carbon monoxide and methane, among others. So by adding to the burden of these gases in the atmosphere, human activities are also increasing the amount of ozone there. Ozone was first discovered in the 1850s, although it was not until the end of the nineteenth century that its chemical composition was firmly established. The first measurements of the concentration of ozone in the air near the ground were made in the Tatra Mountains (now part of Poland) in the 1930s, and gave figures in the range from ten to twenty-five ppb. Modern measurements, from the ground and using balloons, show that there is more tropospheric ozone in the northern hemisphere, near the sources of industrial pollution, than there is in the south. The concentration varies from place to place and time to time, but seems to be increasing at something between 0.5 per cent and four per cent a year.

The lifetime of an ozone molecule in the troposphere is only a few weeks, so the concentration never has time to settle down to a steady global figure in the way that the concentrations of the other greenhouse gases do, but for the sake of having a number to play with researchers often use a notional figure, derived from many observations around the world, of forty ppb for the tropospheric ozone concentration today. With a growth rate of one per cent a year, that would become sixty-four ppb by 2030, an increase on the present day concentration of more than fifty per cent. The pre-industrial concentration may have been about a quarter of the present concentration, or even less.

In spite of the small amounts present in the troposphere, however, ozone is such an efficient absorber of infrared radiation that the increase over the next few decades will contribute a greenhouse warming stronger than the extra warming caused by the build-up of nitrous oxide over the same time, and nearly as strong as the extra warming caused by the continuing build-up of methane.

This highlights the different ways in which the different greenhouse gases contribute to the global warming. There is so much carbon dioxide in the atmosphere already, absorbing most of the radiation there is to absorb in the appropriate part of the infrared spectrum, that even doubling the concentration does not produce a dramatic change in world temperature (it certainly does not double the strength of the carbon dioxide greenhouse effect; technically speaking, at present day levels carbon dioxide absorption in the infrared is proportional to the logarithm of its concentration). Methane and nitrous oxide are less abundant, so in the part of the atmospheric window where they absorb infrared there is still plenty of radiation for each new molecule to absorb, and the amount of heat absorbed is roughly proportional to the square root of the concentration of each of those gases in the atmosphere. But there is so little ozone about to start with, and each molecule is such an efficient infrared absorber (in the band from eight to ten micrometres), that each new molecule added to the air absorbs its maximum possible amount of heat – the amount of radiation absorbed is simply proportional to the concentration of ozone molecules in the air. Halving the amount of ozone in the troposphere would cool the surface of the globe by 0.5°C, even though there is so little there; doubling the present day concentration would warm the world by 0.9°C.

The last family of greenhouse gases that I shall describe in detail are comparably efficient, for the same reasons. Adding a single molecule of one of the CFCs to the atmosphere contributes ten thousand times more to the greenhouse effect than adding a single molecule of carbon dioxide. There were no CFCs on Earth until they were invented at the end of the 1920s and began to be released in large quantities in the 1950s, as a result of human ingenuity; but they have very long lifetimes (some more than a hundred years) and their concentrations are building up rapidly. By 1985, the contribution of CFCs to the greenhouse effect was about eleven per cent of the contribution from anthropogenic carbon dioxide – but the CFC effect had built up over three *decades*, while the carbon dioxide effect had built up over two *centuries*. Looking at the thirty years from 1955 to 1985 alone, CFCs contributed a warming thirty per cent as large as the warming caused by the amount of carbon dioxide released over the same interval. In addition, they are directly responsible for the destruction of ozone in the stratosphere, which will allow more incoming solar energy to penetrate into the troposphere and perhaps to the ground – not, strictly speaking, the greenhouse effect, but an additional source of global warming near the ground.

The chlorofluorocarbon connection

CFCs are very stable molecules, and the two most widely used to date, CFC-11 and CFC-12,* do not seem to react at all with

---

*The number associated with the shorthand name for each CFC indicates, in a standard notation, the number of fluorine, chlorine, hydrogen and carbon atoms in each molecule. CFC-11 is known chemically as trichlorofluoromethane (CCl_3F), and CFC-12 is dichlorofluoromethane (CCl_2F_2); I shall stick with the shorthand names.

anything in the troposphere. Because they are so stable chemically, they are non-poisonous and non-inflammable. It happens that they vaporize at temperatures between about −40°C and 0°C (unless they are under pressure). This combination of properties made them ideal for use as the working fluid in refrigerators (the stuff that gurgles through the pipes in the back of the fridge), and later as the propellants in spray cans – the gases that, under pressure, push out the active ingredient of the aerosol spray. More recently, they have been used to blow the bubbles in many forms of foamed plastic, and to clean grease from computer circuitry.

Because CFCs are so stable, however, they build up in the atmosphere. The gases diffuse around the globe, and have been found (using instruments based on one of Jim Lovelock's inventions) in the troposphere at the South Pole – even though the great majority of releases have occurred in the northern hemisphere. They also diffuse upwards through the atmosphere, and eventually reach the stratosphere. There, ultraviolet radiation from the Sun (which cannot reach the ground, because it is absorbed by stratospheric ozone) breaks CFC molecules apart, releasing, among other things, atoms of chlorine. Chlorine is highly reactive, and destroys ozone in the stratosphere. It is the presence of chlorine from CFCs that causes a 'hole' to appear in the ozone layer over Antarctica each spring, and there is some evidence, from measurements made from satellites and from the ground, that chlorine from CFCs has caused a worldwide depletion of ozone in the stratosphere, by a few per cent of the natural concentration, over the past ten to fifteen years (this ozone 'problem' is discussed in the Appendix). Because of fears about damage to the ozone layer, several nations have now reached an agreement, known as the Montreal Protocol, to limit the release of these gases in future. The agreement is complex, but essentially operates in three stages. First, consumption of the main CFCs is to be frozen, in 1990, at the levels reached in 1986; then, in 1994, consumption will be trimmed by twenty per cent; finally (and subject to further international agreements) consumption will be cut to fifty per cent of the 1986 figure at the end of the 1990s. But even this action will not stop the build-up of CFCs in the troposphere – not only because many other countries have not signed the Montreal agreement, but also because CFCs have such a long lifetime.

On average, a molecule of CFC-11 (chiefly used in blowing plastic foams and in spray cans) takes eighty years to reach the stratosphere and be broken up by ultraviolet radiation; a molecule

144

of CFC-12 (chiefly used in spray cans, refrigeration systems and mobile air conditioning, but also in plastic foams) stays in the air for 170 years before being broken up. Comparing the rate at which the CFCs were being emitted in 1986 with estimates of the rate at which they were being destroyed as they filtered into the stratosphere, researchers calculated that in effect five-sixths of the CFCs released each year stay in the atmosphere. In other words, simply to maintain the present concentration of CFCs in the air would require a cut in production to one-sixth of present figures – about fifteen per cent. In order to reduce the build-up of CFCs in the air, production has to be cut still further – for all practical purposes, to zero. But one of the most astonishing features of the CFC story is how little of these gases is needed to alter the heat balance of the globe and to damage the ozone layer. In 1985, the concentration of CFC-11 in the troposphere was just 0.22 parts per billion, while CFC-12 checked in at 0.38 ppb. In round terms, these concentrations are one millionth of the concentration of carbon dioxide in the air. Even though the concentration of each of these CFCs was increasing at a rate of several per cent of these figures each year in the 1980s, this really is a very small amount of gas to have such a large impact on Gaia.

Together, CFC-11 and CFC-12 now contribute about the same amount of warming to the global greenhouse effect as methane; but since part of the methane effect is natural, while all of the CFCs are man-made, in terms of human impact on the greenhouse effect they rank second only to carbon dioxide. Even if the present rate of release of these compounds were held steady, their concentrations in the atmosphere would continue to increase for at least a hundred years, until a balance was made with the destruction of CFCs in the stratosphere. At constant emission rates, the concentrations would double by 2030.

Just before the Montreal Protocol was agreed, projections of future emissions based on the increasing use of CFCs suggested that even higher figures would be reached in the early twenty-first century. In one calculation of the resulting contribution to the greenhouse effect, Tom Wigley, of the University of East Anglia, concluded that the equivalent of a doubling of the pre-industrial carbon dioxide concentration would occur by 2019 (including effects from methane and nitrous oxide, as well as CFCs). After the Protocol was signed, Wigley went back to his calculations and worked out a new forecast on the assumption that the agreement will be strictly adhered to. Because even reducing emissions of CFCs by fifty per cent (and that not until 1999) only slows the build-up, and there is so much in the air already the change only pushes the date of the effective doubling of the carbon dioxide concentration out to 2027, giving us a breathing space of just eight years. But even that is not the end of the story.

The Montreal Protocol is designed only to protect the ozone layer, and does not address the greenhouse problem. It actually encourages, indirectly, the production and release of other gases, including other CFCs, that are perceived as being less harmful to the stratosphere. One of the compounds that is now being developed and promoted as a substitute for CFC-11 is another chlorofluorocarbon, CFC-113. This has, it is estimated, less potential to harm the ozone layer than the gas it replaces. But it is also *twice* as effective as a greenhouse gas. The greenhouse potential of other CFC substitutes has, in many cases, simply not been calculated. Future revisions of the Protocol may take the greenhouse effect into account, as politicians and the public become more aware of the greenhouse problem. But all this will take time, and meanwhile other greenhouse gases are being added to the atmospheric burden. So existing scenarios for the early twenty-first century that assume a continuing build-up of CFCs are probably reasonably reliable. As far as the influence on global temperature is concerned, if CFC-11 and CFC-12 don't get you, then their substitutes almost certainly will.

The greenhouse effect of CFCs is certainly not in doubt, in spite of their small concentrations in the air. Measurements from balloons flown at an altitude of five kilometres in the atmosphere have actually recorded the infrared radiation from CFCs at higher altitudes, on its way back down to the ground. CFC-12 absorbs outgoing infrared radiation at 10.8 micrometres, and then re-radiates this energy at the same wavelength. Some of the re-radiated energy goes back down to the surface of the Earth, contributing to the greenhouse effect. W.F.J. Evans, a Canadian researcher, measured this downward flow of energy above Palestine, Texas, as 0.05 watts per square metre. Allowing for the effect of all the CFC-12 in the air below five kilometres, this should correspond to a warming at a rate twice as large, 0.1 watts per square metre at the surface – which compares nicely with calculations of the CFC-12 greenhouse effect today, which give a figure of 0.12 watts per square metre. The measurements, published in June 1988, provide a graphic demonstration of the reality of the anthropogenic greenhouse effect, and lend weight to the calculations made by the theorists. Some of these

146

calculations show that in addition to this direct warming effect, the CFCs also allow more heat from the Sun to penetrate through the stratosphere to the surface of the Earth, as the ozone concentration high above our heads is diminished. A pioneering investigation of this additional connection between CFCs and climate was made at NCAR in the mid-1980s, by Veerhabadrhan Ramanathan (now at the University of Chicago) and his colleagues.

Photochemical reactions involving chlorine from CFCs reduce the amount of ozone in the middle and upper parts of the stratosphere, above about thirty kilometres. This allows more ultraviolet radiation to penetrate deeper into the atmosphere. But one of the effects of this is that some of the ultraviolet radiation interacts with molecules of di-atomic oxygen to produce *more* ozone in the lower stratosphere. The change in the amount of ozone overhead, looking up through a column of the atmosphere from the ground, may be quite small, but the ozone has been redistributed at different altitudes. Curiously, *both* effects make the surface of the Earth warmer. The depletion of ozone above thirty kilometres does so simply by letting more energy through into the lower part of the atmosphere. The stratosphere cools, and it does so even more than you might expect, since as well as absorbing incoming solar energy ozone, remember, is an absorber of outgoing infrared energy. The greenhouse effect of the high stratosphere is diminished – but the greenhouse effect of the low stratosphere, where more ozone is produced as a result of these changes, is increased. In the upper stratosphere, the solar effect dominates, but in any case the two cooling effects work together; in the lower stratosphere the warming influence of the greenhouse effect is more important, but since more ultraviolet is being absorbed at these altitudes because of the changes, once again the two effects march in step.

Updating these calculations in 1988, Guy Brasseur, of the Belgian Institute for Space Aeronomy, and Matthew Hitchman, from NCAR, calculated that by 2050 the amount of ozone lost from a column of air overhead in a region near the equator will be only one per cent. This holds over an area containing seventy per cent of the Earth's surface, within latitude bands 45° either side of the equator. To some extent, it gives the lie to scare stories about depletion of stratospheric ozone allowing ultraviolet radiation to penetrate to the ground and give us all skin cancer while killing crops in the field. But because of the changes in the distribution of ozone, the high stratosphere may cool by 15°C on this scenario, while the lower stratosphere and troposphere warm by several degrees. This is far more alarming than the imagined threat from ultraviolet radiation, since it has the potential to upset the global climatic balance.*

Because it absorbs incoming solar energy, the stratosphere is warmer than the top layer of the troposphere (and the upper stratosphere is warmer than the lower stratosphere). Warm air rising upwards from the surface of the Earth as a result of convection (and cooling as it rises) cannot rise into the stratosphere, because the stratosphere is warmer still. The stratosphere acts as a lid on the tropospheric circulation systems of the globe. Simultaneously cooling the stratosphere and warming the troposphere reduces the effectiveness of this lid, and will allow convection systems to extend to higher altitudes, forming clouds further above the surface than today. A real glasshouse, ironically, actually works by suppressing convection – the hot air inside the greenhouse is trapped by the glass, and cannot escape. The global greenhouse effect, by contrast, enhances convection, by making the air near the surface warmer, so that it will rise more vigorously, *and* by taking away the lid on top of the convective region. Will this encourage the formation of the kind of clouds that, on balance, reflect away incoming solar heat, thereby alleviating the greenhouse problem? Or will it cause the spread of clouds that trap surface radiation on the way out into space (like the clouds that can hold in heat at night and prevent frost in winter), thereby making the problem worse? Nobody can say, just as nobody can say what effect the change in convection will have on wind and weather patterns worldwide. But what *is* clear is that the anthropogenic greenhouse effect is *not* simply about an increase in average temperatures around the globe. It is important to keep that in mind, as we look now at just how large those temperature increases themselves are likely to be, in the first quarter of the twenty-first century.

---

*Only to some extent, because the calculations do not allow for the kind of chemical reactions that cause the hole in the sky over Antarctica and which may also occur, about one-tenth as effectively, at other latitudes. This is discussed in the Appendix. Measurements from Arosa, in Switzerland, already show a decrease in total ozone of between one and two per cent between 1960 and 1980, while observations from Mauna Loa suggest a decline of as much as three per cent during the 1970s. These are suggestive, not definitive, figures – but one thing they suggest is that CFCs are having a bigger influence on stratospheric ozone (ninety per cent of the ozone is in the stratosphere) than straightforward calculations of the kind used by Brasseur and Hitchman indicate.

All things considered

The calculation by Ramanathan and the NCAR team of the connection between CFCs, stratospheric ozone and temperature changes at different altitudes through the atmosphere was actually part of a much bigger study of the way in which just about every known greenhouse gas would contribute to the global warming over the next few decades. There are many more potential greenhouse gases than I have even hinted at so far. These include other CFCs, and some close relations of the CFCs known as halocarbons. Carbon tetrachloride, for example, has a lifetime in the atmosphere of about forty years, and is released as a result of industrial activity. In some parts of the world, it is still used as a cleaning fluid in so-called 'dry cleaning' processes; it is also an intermediate step in the production of CFCs, and it is an effective greenhouse gas. Other greenhouse gases are unrelated to CFCs; these include sulphur dioxide. In their study published in 1985, the NCAR researchers considered the build-up of forty trace gases in the atmosphere, including calculations of anthropogenic emissions and of natural greenhouse gases. About twenty of these substances absorb strongly at wavelengths between seven and thirteen micrometres. The NCAR team allowed for the overlapping effects of absorption by these gases in the atmospheric infrared window, and calculated how the combined effects would be likely to warm the world over the fifty years from 1980 to 2030; they found that carbon dioxide is contributing less than half of the total effect. But more than ninety per cent of the present warming is caused by a combination of carbon dioxide, methane, nitrous oxide, the two main CFCs and changes in ozone concentrations. So for the time being it is reasonable to consider the other gases as also-rans, while keeping an eye on them in case emissions start to increase as a result of some new industrial process or substitution for CFC-11 or CFC-12.

The price that has to be paid for going into such detail in calculating so many different contributions to the greenhouse effect is that no available computer can do all this and calculate the subtle interactions portrayed by a GCM at the same time. Ramanathan and his colleagues used a one-dimensional model, which means that surface temperatures are given as a global average, as if all of the surface of the world were at the same temperature, but temperature variations with altitude, up through the atmosphere, are computed. So they are able to include the additional warming effect caused by depletion of ozone in the stratosphere, as well as the greenhouse effect of tropospheric ozone – and all the other gases – but *not* the feedback caused by, for example, changes in ice cover at high latitudes.

When such a one-dimensional model is run to calculate the warming effect of doubling only the amount of carbon dioxide in the atmosphere, it typically gives a figure of about 2°C, perhaps lower, for the global average increase. When a GCM, with all the feedbacks mentioned earlier, is used to calculate the warming that will result from doubling the carbon dioxide concentration, it generally gives a figure of 3.5°C for the global average, or even higher. So it is a reliable rule of thumb that forecasts from one-dimensional models should be doubled to give a more accurate guide to what is likely to happen in the real world. Equally, the temperatures projected for specific years in scenarios from GCMs for a doubling of carbon dioxide alone should also be doubled (at least), because it turns out that over the appropriate timescale carbon dioxide contributes no more than half the total anthropogenic warming. As a rule of thumb, *all* the simple forecasts of temperature increases over the next forty to fifty years tell only half the story.

Without allowing for this, the NCAR study showed that the surface warming between 1980 and 2030 caused by the build-up of all the trace gases considered would be 1.54°C. Carbon dioxide contributes just 0.71°C to this figure. CFC-11 and CFC-12 together contribute 0.36°C to the direct greenhouse warming, and a further 0.08°C due to the change in stratospheric ozone. This makes their overall influence 0.44°C, about sixty per cent as large as the carbon dioxide effect.* Methane contributes 0.14°C, nitrous oxide 0.1°C, and tropospheric ozone 0.06°C. The same computer model suggests that the warming that had already taken place before 1980, as a result of anthropogenic contributions to the greenhouse effect, was about 0.8°C, with 0.5°C due to carbon dioxide and the rest to other trace gases. So the total warming forecast, from pre-industrial times to 2030, is about 2.3°C, which is almost exactly the same as the forecast for a doubling of carbon dioxide alone using the same model. Wigley's more recent calculations, using slightly more up-to-date emission figures, suggest that this equivalent of

*Remember, this is a *scenario*. If we release less CFCs because we are alarmed at the implications, the world may warm less (provided the substitutes are not even more effective greenhouse gases). If so, that doesn't mean the scenario was wrong, but that it was a useful guide which altered our attitudes.

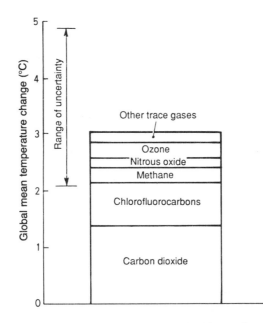

Figure 6.6
The likely overall warming of the world caused by a build-up of greenhouse gases released as a result of human activities between 1980 and 2030 (including allowance for feedback effects).
(Adapted from information in V. Ramanathan et al., Journal of Geophysical Research, *volume 90 pages 5547 to 5566, 1985)*

a carbon dioxide doubling will be reached in 2027 *even if* the Montreal Protocol takes full effect, and substitutes for CFCs are not themselves greenhouse gases; otherwise, the doubling *could* be reached by 2020.

The simplest assumption is that this carbon dioxide equivalent is a good guide, and we can take over the figure for a temperature increase for a carbon dioxide doubling from the GCMs – 3.5°C, or more. Then, we can translate the intermediate steps, including the 0.8°C rise from pre-industrial times up to the 1980s, by multiplying by the number required to convert one-dimensional calculations of carbon dioxide alone into their GCM equivalents. There is some uncertainty, because different GCMs give slightly different forecasts. In the SCOPE 29 report, a conversion factor of 1.7 is used, which probably errs on the cautious side, but that report also suggests that the factor might be as high as 2.5. I would settle for a factor of 2; but on the basis of the SCOPE figure of 1.7, the warming prior to 1980 in the real world should have been about 1.4°C. Calculations of the thermal inertia of the oceans suggest that only about forty per cent of this warming, perhaps 0.6°C, should have shown up yet in the record of average global air temperatures. This is exactly in line with the temperature record of the past century or so, and suggests that there is *already* a further 1°C rise, almost twice as big as the change over the past hundred years, stored up in the system and working its way through to us. If carbon dioxide alone had been responsible for these changes, the concentration in the atmosphere needed to produce the same warming by 1980 would have been 380 ppm; the carbon dioxide equivalent of the total greenhouse effect forecast for 2030 is 590 ppm, comfortably (if that is the right word) more than double the pre-industrial concentration.

The thermal inertia effect may continue to slow the warming due to the continuing emission of greenhouse gases, but this is not likely to be a great help by the time this kind of carbon dioxide equivalent has been reached. When climatologists talk of a doubling of the carbon dioxide equivalent by, say, 2030 implying a rise in global mean temperatures of 3.5°C, it is understood that they mean the world would warm that much *once equilibrium had been established*. Establishing equilibrium might take a decade or so, giving us a little more breathing space; then again, as the concentration of greenhouse gases may still be rising in the decades after 2030, equilibrium may not really be established at all, and temperatures would then rise faster than if emissions stopped at the level equivalent to a doubling of carbon dioxide. The delay itself is likely to decrease as the weather machine is driven further away from equilibrium by the build-up of greenhouse gases – the harder you push it, the faster it will move. Forecasting just how much breathing space, if any, the thermal inertia effect will give us as we move into the twenty-first century is almost impossible, although some researchers do try to take account of this lag in their calculations of the actual warming we are likely to experience in the 1990s. Other figures, such as the 3.5°C by 2030 scenario, should, strictly speaking, be read as, for example, 'a doubling of the carbon dioxide equivalent by 2030 will *commit the world to* a warming of 3.5°C'; the temperature in the year the doubling is reached will not yet be quite that high, but will be rising rapidly towards that figure.

But don't let that lull you into a false sense of security. For this kind of carbon dioxide concentration, equivalent to what we will be experiencing within four decades, many GCMs forecast a warming, once equilibrium has been established, of four to five degrees Celsius, compared with pre-industrial averages. We might well hit a warming of 3°C or more in 2030 even allowing for thermal inertia, and with more warming still to come even if emissions of greenhouse gases were magically turned off. The small increase from the nineteenth century to 1980 is already being swamped. Global average temperatures are likely to be committed to rise by 5°C from their *present* levels by the middle of the next century.* Even allowing for the lag caused by the time it takes for ocean waters to warm up, the warming we have experienced since the nineteenth century is quite insignificant compared with what we can expect over the next few decades. And even if emissions of all greenhouse gases ceased tomorrow, the average temperature of the world would rise by a degree or more over the next twenty or thirty years, as the climate system adjusted to the new equilibrium. It is the speed of the likely greenhouse warming, as well as its size, that makes this altogether too much of a good thing. Such a dramatic change, bigger than the change from an ice age to an interglacial, but occurring in tens of years, not thousands, could switch the weather systems of the world into a completely different pattern.

*A supreme optimist might halve these figures, in the hope that the lowest end of the range of GCM forecasts is accurate. In other words, the *optimistic* view is that average temperatures will rise by 'only' 2.5°C by 2050.

SEVEN

All at Sea

The oceans are the key to understanding changes in global climate – but, to mix the metaphor, they are also the joker in the pack. Most of our planet is covered by water, and interactions between the oceans and the atmosphere establish the balance of natural greenhouse gases that has maintained temperatures on Earth comfortably above those of the airless Moon, but comfortably below those of hothouse Venus. The seas also smooth out the distribution of heat across the planet, transporting warmth from the tropics, where solar heating is strongest, to higher latitudes, where the direct heat of the Sun is weaker. Unfortunately, we do not understand the workings of the oceans, and of the air/sea interface, well enough to predict exactly how these patterns will change as the world warms. But there is a disturbing possibility that any changes that do occur will be sudden, a switch in the mode of oceanic circulation from one pattern to another, rather than gradual.

Today, the atmosphere and the oceans each carry about the same amount of heat from the equator to the poles, but they do it in different ways. In the atmosphere, from the equator out to about 30° latitude in each hemisphere, the dominant feature is simple convection, with hot air rising near the equator, cooling and moving outward high in the troposphere, and sinking down to the surface again to become a cool wind blowing into the equator from higher latitudes. This circulation pattern produces the reliable trade winds that were of such prime importance to shipping in the days of sail. Further away from the equator, heat is redistributed in a more messy fashion by the weather systems – the highs and lows – which are familiar to anyone who lives in temperate latitudes. The weather systems are actually eddies, like the swirling patterns formed in flowing water when it moves past an obstruction. At these latitudes, sometimes the wind blows one way, at other times it blows from a different direction. But by and large the effect is the same – air blowing towards the poles is warm, while air blowing from the polar regions is cold.

The way oceans transport heat from the equator to the poles is different for two reasons. First, the oceans are heated from the top, while the atmosphere is heated from the bottom. Warm ocean water heated by the Sun cannot rise because it is already at the top of the ocean. So although there is convection in the oceans, it is, in a sense, upside down convection compared with what we are used to in the atmosphere. Atmospheric convection is driven by hot air rising at the equator; but oceanic convection is driven by cold water sinking at high latitudes. Just as the warm air rising up pushes other air out of the way and sets the atmospheric convection circulating, so the sinking cold water at high latitudes pushes other water out of the way, eventually ensuring that water rises to the surface in the tropics and is warmed by the Sun as it begins to move out towards the poles.

But even then the ocean currents cannot flow as freely as the winds blow, and the second difference between oceanic and atmospheric currents comes into play. There are huge land masses which divide the oceans of the world into two main basins in the northern hemisphere (the Atlantic/Arctic and the Pacific) and three in the south (Atlantic, Indian and Pacific). Although the southern ocean basins do join at high latitudes to form the Southern Ocean, surrounding Antarctica, the way most of the waters of the world are confined by continental masses forces the currents to flow around the basins in roughly circular patterns, called gyres (*Figure 7.1*). The path followed by such an oceanic current depends partly on the difference in temperature between the equator and the poles, partly on the effect of the Earth's rotation, and partly on the shape of the ocean basin itself. The Earth's rotation (from west to east) spreads out currents flowing along the eastern side of a basin, but piles up the currents into strong, relatively narrow streams in the western sides of the ocean basins – the western land mass is always moving towards the water, while the eastern continents are moving away from it (which is why sea level at the Pacific end of the Panama Canal is lower than sea level at the Caribbean end). The Gulf Stream which runs northward up the western side of the North Atlantic is the archetypal example of all these processes at work. It carries thirty million cubic metres of water every second

155

through the Florida Straits, and that water is 8°C warmer than the water that returns south as the gyre completes its circuit of the basin. The rate at which heat is being transported northward is more than a million billion watts.

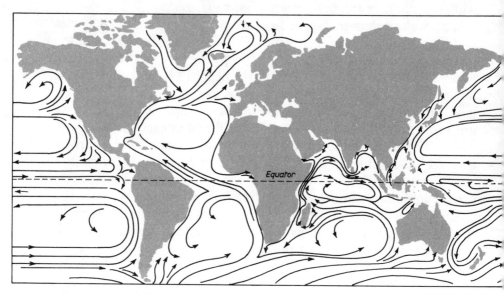

Figure 7.1
The pattern of ocean surface currents.

The Atlantic is, in fact, the only basin which provides heat for the Arctic Ocean in this way; the northern Pacific is almost completely blocked off at the Bering Strait, where the Soviet Union and the United States very nearly have a land frontier (the strait was, indeed, dry land during the recent series of ice ages, when sea level was lower). So in the Pacific Ocean, equatorial heat cannot reach up to the highest northern latitudes, and most of the warmth flows southward, linking up with the warm southward flow from the Indian Ocean (which is completely blocked to the north). Then it splits into two parts: a current that flows into the Southern Ocean and eventually transports heat to high southern latitudes, and a current which flows around southern Africa and into the Atlantic, where it moves northward and eventually helps to make the warm currents flowing poleward in the *North* Atlantic stronger.

In the south, meantime, the main current in the Southern Ocean flows in an almost circular path around Antarctica, as the Circumpolar Current. Although this provides a barrier to the

156

direct flow of warm water southward, heat is carried across the Circumpolar Current in the form of eddies, rather like the swirling low pressure systems that transport atmospheric warmth northward at the latitudes of the United States and Europe. About 300,000 billion watts of heat is transported in this way – significantly less than the equivalent northward flow, because of the way the shape of the ocean basins forces the currents to flow.

Over the long span of geological time, the geographical patterns of the land masses have changed as continents have drifted about the globe. This has altered the circulation patterns of the oceans, and has caused dramatic changes in climate – we live in an epoch when there is ice at both poles, but when the dinosaurs dominated the Earth there may only have been ice on high mountains. Then ocean currents carrying warm water could reach both polar regions largely unobstructed.

Clearly, the pattern of ocean currents today fits in with the geographical pattern of the continents today. But is it the *only* pattern of ocean currents that fits the present day geography of the globe? One person who thinks it is not, and who suspects we may be in for a 'flip' of the current system (and therefore the climate) into a new pattern is Wallace Broecker. It is, he says, just a hunch; but as we have seen, Broecker's hunches have an uncanny knack of coming true.

Unpleasant surprises ahead?

The climate of our planet is determined by a combination of factors, known as boundary conditions: the geography of the globe, which determines ocean circulation patterns; the amount of heat reaching the Earth from the Sun; the amount of carbon dioxide in the air; and so on. When one of the boundary conditions changes slowly but steadily, you might expect climate to change in the same way – with the Earth, for example, steadily getting warmer if the output of heat from the Sun steadily increases. But two researchers based in Texas, Thomas Crowley and Gerald North, recently showed that this is far from always being the case. Sometimes slowly changing boundary conditions produce a sudden jump in climate from one state to another, a discontinuous change as a result of a 'last straw' effect. And when that happens, it doesn't necessarily follow that slowly restoring the boundary conditions to their previous values will cause the climate to flip back into the state it used to be in.

The Texas researchers carried out their study (not the only one of its kind, but particularly relevant to the present story) to show how sudden changes in climate millions of years ago might have contributed to ecological disasters when many species of plants and animals disappeared from the face of the Earth in a short period of time. Fossil evidence points to five such 'major extinctions' during the past six hundred million years. But their work may also be relevant to the world of the twenty-first century.

The jumps happen because the climatic systems of the Earth may be able to exist in two (or more) stable states for a single set of boundary conditions. To keep things simple, imagine that everything else stays the same, but the carbon dioxide concentration of the atmosphere slowly increases. There is a particular range of values for the greenhouse effect where the Earth can exist quite happily either with or without an icecap over the Arctic Ocean. If the ice is present, then it reflects away enough incoming solar energy to keep the pole cool in spite of the greenhouse effect, and thereby to maintain itself. If the ice is not there then the polar sea is warmer, because it absorbs more solar energy than the shiny ice does, and the ice cannot form.

If the world starts in this range of possible states, with an icecap, and the greenhouse effect steadily increases as carbon dioxide builds up in the atmosphere, there will come a time when the warming effect is so strong that the ice begins to melt back rapidly, as its cooling influence is overwhelmed. As it does so, more dark water is exposed, the polar sea warms once more, and in a very short time the positive feedback removes all the ice. The whole northern hemisphere will be warmer as a result, by several degrees Celsius, even though the greenhouse effect has only increased by a tiny amount. But now, if by magic the carbon dioxide content of the atmosphere is steadily reduced once more, the ice will not come back immediately. Instead, the world must cool to a point where ice begins to grow near the pole in spite of the dark ocean's ability to absorb heat. Then the ice reflects away solar heat, the polar region cools still more, and the icecap soon re-forms – but none of this happens until the carbon dioxide concentration has fallen far below the critical value at which the icecap disappeared when the concentration was *increasing*.

When the two Texas researchers put some numbers into their model calculations of this kind of process, they found that at the critical point the flip can occur for a temperature change equivalent to a variation of just 0.0002 per cent in the Sun's output. This is less than one ten-thousandth of the size of the effect associated with changes in the Earth's orbit, the Milankovitch cycles. It is so small that it suggests that human contributions to the greenhouse effect could well have a dramatic influence on climate in the near future, when some critical threshold is reached and passed. Unfortunately, because the real world is more complicated than the computer models, the models cannot, as yet, tell us exactly when such a threshold will be reached, or which part of the climate system is likely to flip first. Melting the north polar icecap may not be on the agenda for another fifty years or so, as we shall see in Chapter Nine. But ocean circulation patterns may respond to the steadily growing greenhouse effect even before then, if Broecker's instinct is correct.

The analogy Broecker himself makes is with the onset of the monsoon in India each year. In winter, the Tibetan Plateau is cold and air spills down from the plateau to the south and west, holding moist air from the oceans at bay. But in summer, the plateau steadily warms up as the Sun rises higher in the sky. At a critical moment, hot air begins to rise strongly from the plateau, sucking in moist, warm air from the Indian Ocean to replace it and bringing the monsoon rains. The cause of the change is the steady increase in solar heating over a matter of months; but the change itself, from one circulation pattern to a completely different pattern, happens literally between one day and the next. Can similar changes take place in the oceans?

Certainly, *something* can make the climate of the Earth jump from one state to another, at least during an ice age. The ice core records from Greenland show that within the most recent ice age there were several occasions when the temperature suddenly jumped by about 6°C, with a fivefold change in the amount of dust in the air and a twenty per cent change in the carbon dioxide concentration. These natural changes took place in the span of a few hundred years, with more dust and less carbon dioxide in the air when the world was colder.* Broecker links these leaps in climate with flips of the ocean currents from one pattern to another, changes in what he calls a global conveyor belt carrying heat around the globe.

At present, the Earth's climate system works to the benefit of northern Europe. In the North Atlantic, as I have mentioned,

*The unnatural anthropogenic greenhouse effect will produce changes as big as this *even more rapidly*, in less than a century if no action is taken to slow it.

warm water is arriving as surface currents. Some of this warm water can be traced all the way back to the Indian and Pacific Oceans. The water gives up its heat to the cold winds blowing across the North Atlantic from Canada, and sinks, starting the upside down convection of the ocean current system. The amount of heat it gives up is nearly one third as much, each year, as the region receives from the Sun; this mostly benefits Europe, which is downwind of the ocean. Cold, deep currents carry the circulation back into the Indian and Pacific Oceans, where the water rises once more (pushed by the flow behind it) and warms as it begins its long journey back to the North Atlantic (*Figure 7.2*). The pattern is maintained, Broecker points out, by salt. Because the conveyor belt operates in this way, the North Atlantic is warmer than the North Pacific, so there is proportionately more evaporation from the ocean's surface there. If more water evaporates, then the water that is left behind must become more salty, and salty water is more dense, so it will sink. The resulting cold, deep flow starts out as the North Atlantic Deep Water Current, an oceanic 'river' that carries a flow of water twenty times greater than the flow of all the rivers of the world put together; eventually, it carries salt into the Pacific to compensate for the imbalance. The whole pattern is self-stabilizing; in Broecker's words, 'excess evaporation causes the deep current; the deep current causes excess evaporation'. But if the conveyor belt were even temporarily shut down, then the land around the North Atlantic would cool by about 6°C – exactly the size of the change in temperature associated with climate flips during the latest ice age. 'The many leaps in climate seen in the glacial [ice age] part of the Greenland ice core probably represent flips of the system back and forth between two self-stabilizing modes of operation.'

We can imagine that the pattern might reverse, with the North Pacific being the beneficiary of warm surface water, and the North Atlantic becoming the region where cold, deep water returns and rises to the surface. But just turning the system off for a while would explain the climate flips of the latest ice age. One way to turn the system off would be to flood the North Atlantic with fresh water, floating on top of the salty sea, as may have happened during the dramatic cooling that occurred some eleven thousand years ago, after the great ice sheets had begun to retreat. But the point of Broecker's study is not to spell out exactly how past flips have occurred, but to stress that flips are possible.

Whatever the causes of these changes, the links with the amount of carbon dioxide in the air are clear, since the descending cold

160

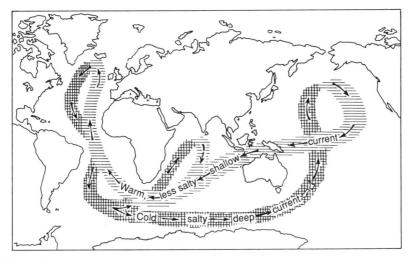

Figure 7.2
The oceanic 'conveyor belt' that transfers warm, less salty water from the Pacific to the Atlantic as a shallow current, and returns cold, more salty water from the Atlantic to the Pacific as a deep current. If the conveyor slows, Europe freezes. (Adapted from a figure presented by Wallace Broecker to the December 1987 meeting of the American Geophysical Union)

water carries with it a burden of dissolved carbon dioxide from the atmosphere. One message from this is that if the conveyor belt does turn off again, the oceans may absorb less carbon dioxide than they do at present, perhaps causing a sudden upward jump in the atmospheric concentration of twenty per cent, over and above the increase caused by human activities. The good news is that the system seems to have worked very steadily for the past ten thousand years, during the present interglacial, with none of the flips that seem to be characteristic of ice ages proper. But we are now forcing the Earth into a super-interglacial, an unnatural state in which a geographical pattern and ocean circulation system that ought to be associated with cold and polar icecaps is, increasingly, being forced to accommodate a warm world. There is no analogy from the past to draw on to help us guess how the climate will behave under those conditions, and no way we can predict how the climate might jump as a result. But Broecker is convinced that change, when it happens, will occur abruptly, and will probably involve changes in the oceanic conveyor belt that transports heat, salt, and indeed carbon dioxide, around the globe. If we are lucky, dramatic changes of the kind he envisages may still be decades away. But the importance of ocean currents in establishing weather

161

patterns worldwide is clearly shown by the record-breaking heat of
the 1980s, and its links with the Pacific Ocean pattern known as
El Niño.

Children of the sea

The climatologists' El Niño is not quite the same as the local-
ized phenomenon which was given the name by South American
fishermen. They noticed that about every four years or so the
water near the South American coast of the equatorial Pacific
becomes unusually warm. Because the phenomenon often happens
in December, they gave it the name El Niño, which literally means
'the little boy', but which specifically refers to the Christ child. To
the climatologists, El Niño is a pattern of temperatures over a large
area of the Pacific Ocean, with warm water in the eastern Pacific
and cold water in the west. Normally, the surface layer of water
in the eastern Pacific is colder than the surface water to the west.
Prevailing winds, driven by the rotation of the Earth, push warm
surface water westward, allowing cold water to rise up from the
depths along the Pacific coast of South America. Every few years,
however, these prevailing winds weaken, and this allows a warm
patch of water to develop in the eastern Pacific. Where the ocean
is warm, air rises up vigorously from the surface, sucking in winds
from the west to replace it. These winds then push the warm
surface water of the Pacific towards the east, piling it up above
the cold, deep water, and helping to maintain the El Niño in a
positive feedback. The pattern is self-sustaining, for a while, but
after a couple of years it breaks down and may be replaced by the
reverse effect. Then the eastern Pacific becomes even colder than
normal, while the western part of the ocean warms. Winds once again
blow towards the region of warm ocean, now in the west, piling up
warm surface water there and rising in a convective column above
it. Once again, the pattern is self-sustaining; but once again, in
practice it only lasts for a year or two before breaking down. This
reverse effect has at times been called the 'anti-El Niño'; but since
to a Spanish speaker this means roughly 'anti-Christ child', recently
climatologists have begun to use instead the term 'La Niña' – 'little
girl' being the opposite of 'little boy'.

When the temperature pattern over the oceans reverses, the winds
also reverse, a phenomenon known to meteorologists as the Southern Oscillation – which gives the whole interconnected pattern of

oceanic and atmospheric circulation changes yet another name, El Niño/Southern Oscillation, usually abbreviated to ENSO. Whatever names the phenomena go by, they are of immense importance to the climate of the region around the Pacific, and perhaps to the world at large. The pattern of winds associated with an El Niño, for example, brings drought to Australia. There is no problem in explaining why either an El Niño or a La Niña pattern should persist once it forms, but it is only recently that researchers have found a likely explanation for why the pattern should break down and then reverse, on average, about every three to five years. The idea goes back to the 1970s, but was picked up and developed in the late 1980s by several different teams, using improved computer models.

The basic model contrasts two sorts of wind-driven disturbance in the ocean. The tropical ocean is simply represented, in the computers, as a layer of warm, light water resting on a much deeper layer of cold, dense water. Near the equator, where winds blow almost along lines of latitude ('zonal' winds), surface water is piled up either in the east or the west depending on which way the wind is blowing. Waves produced by the effect of these zonal winds are known as Kelvin waves, and they can cross the entire Pacific basin in two or three months.

Beyond about three degrees of latitude from the equator, however, the winds are distorted, because of the shape of the Earth and its rotation, into a more circular pattern. These curving winds have a different effect on the surface layer of the ocean. Where the winds are anticyclonic (clockwise in the northern hemisphere; anticlockwise in the south), the 'curl' forces push surface water into the centre of the system and deepen the warm layer. Where the circulation is in the opposite sense (cyclonic), there are diverging forces at the surface and the warm layer is thinned. The disturbances that are produced as a result are called Rossby waves, and they travel at speeds of between twenty-five and eighty-five kilometres a day, much more slowly than Kelvin waves. They also travel, always, from east to west across the Pacific Ocean (this essential asymmetry is linked with the rotation of the Earth – because the Earth is rotating, the ocean 'knows' which way north is, and the difference between east and west). During an El Niño, with the sea level high in the east, the Rossby waves carry depressions in sea level slowly back to the western Pacific; during a La Niña, the pattern is reversed. The 'information' about sea level carried by the Rossby waves is always the opposite of the actual state of the eastern Pacific. But it takes a long time for the message to get through. Near the equator, a Rossby wave takes about nine months to cross the Ocean; at 12° latitude, it may take four years. Eventually, however, the negative signal from the east arrives at the curving boundary of the western Pacific where, because of the shape of the coastline, the waves are reflected towards the equator, and become Kelvin waves, *still carrying the same negative signal.* Like all Kelvin waves, they scoot eastward along the equator at a rate of about 250 kilometres a day, arriving in the east with their negative message in a matter of weeks, and cancelling out the conditions which have been maintaining the El Niño.

On the new picture of ENSO events, when a pattern of cool and warm patches of surface water develops, forming a sea surface temperature (or SST) anomaly, the associated changes in wind patterns reinforce the anomaly in the short term – for example, strengthening El Niño. But at the same time Rossby waves are initiated that must in the space of a few years (because of the shape of the Pacific basin) wipe out (and then even reverse) the pattern of SST anomalies. The timescale for this interaction with Rossby waves produces the three to five year delay needed for negative feedback to explain the ENSO cycles. And if an El Niño, for example, lasts for two years, then there will be two years' worth of the opposite kind of Rossby waves still marching around the Pacific and helping to initiate La Niña conditions after the El Niño ends.

The key evidence in support of this explanation of ENSO events came from two computer studies. In one, a computer model simulating this kind of ocean-atmosphere interaction was run for thirty-five model 'years' (it takes forty seconds of supercomputer time to simulate one day of model time; more than 140 hours to simulate thirty-five years), and produced quasi-periodic oscillations with periods of four to six years. In a second simulation, numbers corresponding to the real pattern of winds and temperatures over the Pacific in 1985 were fed into a computer model, and used to forecast how the ENSO would develop. This led to the successful forecast, nine months in advance, of the El Niño that began in 1986, and which was associated with the worldwide pattern of meteorological events that made 1987 hotter than any previous year on record.

Global connections

During the 1980s, the world as a whole warmed considerably. There were repeated El Niño events, but no La Niña until the end of 1988. And there were recurring droughts in the Sahel region of Africa and Ethiopia, while the Indian monsoon was weak. All of these events may be connected, both to each other and to the greenhouse effect. The connections began to become clear in 1987, when researchers at the UK Meteorological Office established that analysis of sea surface temperatures around the globe provides a good guide to the amount of rain that is likely to fall in the Sahel, a region which stretches along the southern border of the Sahara desert from Senegal to Ethiopia.

Broadly speaking, if the seas in the southern hemisphere are warmer than normal during the *northern* hemisphere spring, then the Sahel can expect low rainfall during the wet season, in July and August. The pattern was first spotted in 1984, and led to successful predictions of rainfall in the region in 1986 and 1987. The key to the effect may lie in the northward surface current of the Atlantic Ocean, the main route which transfers heat from the southern hemisphere to the northern hemisphere. When the current is sluggish, the anomalous temperature difference between the two hemispheres develops. Nobody knows why the current should have been sluggish during the 1980s. The winds that drive the current seem to have been weaker, though, and this weakening of the winds may itself be associated with atmospheric circulation changes caused by the greenhouse effect. But in any case, as we have seen, this current is part of Broecker's oceanic conveyor belt, which can be traced back to warm water from the Indian and Pacific Oceans. Indeed, sea surface temperature patterns from all three major oceans combined produce an even better guide to rainfall in the Sahel than Atlantic SST measurements alone, emphasizing the importance of interconnections between the oceans.

El Niño itself seems to have a direct influence on rainfall in many parts of the world. When El Niño events are at their height, the zone of lowest atmospheric pressure that normally sits over Indonesia shifts to the west and there is a change in rainfall patterns around the Pacific, generally leading to a rise in temperatures and causing drought in many places. I have already mentioned Australian droughts; in 1983, the pattern of ENSO winds associated with a strong El Niño that began in 1982 (the strongest since 1877) caused massive drought and forest fires in

Australia. There is also a connection between El Niño and droughts in northeastern Brazil and in Mozambique, and rainfall variations in coastal Kenya.* Most dramatic of all, out of the past twenty-six El Niño years, twenty-one have coincided with failure of the Indian monsoon.

These failures, in particular, are also linked with very warm conditions over the Indian Ocean. During a normal monsoon, warm, moist air moves in from the sea and rises over land, forming clouds and rain as it cools. But when the ocean temperature exceeds 28°C, the warm air rises very strongly *over the sea*, forming storms that drop most of their moisture as rain before they move off over the land. There is a clear implication here that the greenhouse effect, by warming the world, may make failures of the Indian monsoon more common. But why else should the Indian Ocean be warm? Remember that the surface currents there are part of the same global conveyor belt that carries heat, ultimately, into the North Atlantic. If the conveyor belt is moving more slowly, explaining why the South Atlantic is 'anomalously' warm, then it seems reasonable to expect that the Indian Ocean surface currents, further back along the same conveyor, will also be more sluggish and have more time to warm up in the heat of the Sun.

Nobody knows if the strength of recent El Niño events is related to the build-up of carbon dioxide in the air and the long-term rise in surface temperatures. But it is clear that the record-breaking warmth of the 1980s is itself related to the strength of recent El Niño events. The intense El Niño of 1982–83 was not followed immediately by a La Niña, but by a handful of 'normal' years before the onset of another El Niño in 1986. This event was linked with astonishingly warm conditions in Canada, for example, where average temperatures in the winter of 1986–87 were as much as 9°C above normal ('normal', in meteorological jargon, usually refers to the averages for the period from 1950 to 1970). In

*Drought and forest fires, of course, contribute an extra burden of carbon dioxide to the atmosphere, an estimated four thousand million tonnes of carbon as carbon dioxide as a result of the El Niño of 1982–83. At the conference on the Gaia theory held in San Diego in 1988, Richard Gammon, of the US National Oceanic and Atmospheric Administration, pointed out that *if* a warmer world is more prone to El Niño events, then this could act as a 'ratcheting' effect, reinforcing the global warming. Somewhat dramatically, he linked this with the growing size of seasonal variations in the amount of carbon dioxide in the atmosphere, and said, 'it is possible that the northern hemisphere is now a net respirer of carbon. Maybe the dieback has begun'.

1987–88, the anomalies were about half as large, but still sufficient to wreak havoc with a snow-starved Winter Olympics in Edmonton. Worldwide, 1987 and 1988 were the two warmest years on record.

One of the peculiarities of the 1980s, however, is that there were no La Niñas, up until 1988, to counteract the warming influence of the El Niños. In the past, El Niño and La Niña events have occurred with about equal frequency, and whereas El Niños are associated with years of higher than average temperatures (especially near the equator), La Niñas are associated with cool years. The last La Niña prior to 1988 was in 1975, but at the end of 1988, following the two hottest years on record, a distinct La Niña pattern had at last emerged in the distribution of Pacific SSTs. There has been no comparable gap between La Niña events since records began in the last quarter of the nineteenth century. At a time when the world is warming anyway, because of the greenhouse effect, any years that are warmer than average for some other reason (such as El Niño years) will stand out even more, because the average itself has gone up. So how much of the record-breaking heat of the 1980s was due to the greenhouse effect, and how much to the El Niños?

The events of the next few years may tell us. The important point is not that the El Niño years of the 1980s were hotter than the non-El Niño years of the same decade, but that they were hotter even than the El Niño years of previous decades. Something else – the greenhouse effect – is *also* making the world warmer. La Niña generally has the opposite influence to El Niño in affected areas, bringing wetter monsoon conditions to India, more rain to the Sahel, cold winters to western and central Canada, and so on. During La Niñas, the normal winds and currents across the Pacific get stronger, not weaker, and the associated changes in world weather patterns in 1988 were linked with floods in southern China, heavy rains in eastern Australia, and flooding in Bangladesh and the Sudan. One of the most important features of the whole ENSO system, however, is that it affects temperatures primarily in the equatorial region. The greenhouse effect, on the other hand, should produce a more pronounced warming at high latitudes. The fact that the record-breaking heat of 1987 and 1988 was linked with equatorial as well as high latitude warming does suggest that El Niño played a big part. But if the world is getting warmer anyway, La Niñas may well be able to do no more than return temperatures, temporarily, to the 'normal' levels of the 1960s and 1970s. The last time that average global temperatures were actually below the mean for the period from 1940 to 1960 was in 1976 – the year after the last La Niña before the 1988 event began.* But the record for global warmth, prior to 1987, was in 1983 – the year after the strongest El Niño of the twentieth century began. If the global warming seems to level off, or even go into reverse, in 1989–90, that will not mean that we can ignore the greenhouse effect after all; it is simply that one of the many natural fluctuations in climate has temporarily given us a breathing space. But if temperatures stay high even with the influence of La Niña, we can be sure that the greenhouse effect is at work with full force. And either way, when El Niño returns in the 1990s, as it must when the Rossby waves complete their journey across the Pacific, record-breaking temperatures even higher than those of 1987 and 1988 will assuredly return with it. A couple of years of '1950s' temperatures would not mean that the warming of the past thirty years has been reversed overnight, but would, even if they occur, simply represent a short-term downward blip in the rising temperature trend.

La Niña may also, for a couple of years, help to offset the greenhouse effect by stimulating plant growth in regions that benefit from the rain it brings – the reverse of Richard Gammon's gloomy 'dieback' scenario. More speculatively, it may increase the uptake of carbon dioxide by the oceans. La Niña cools between ten and fifteen per cent of the entire oceanic surface of the world, and conventional wisdom has it that carbon dioxide dissolves more effectively in colder water. But such suggestions should be taken with caution. Oceanographers have recently begun to appreciate that they know less than they thought they did about the uptake of carbon dioxide by the oceans, and that a great deal more work needs to be done before they can forecast with confidence how changes in temperature, or sunlight, might alter the efficiency of this carbon dioxide sink.

Problems with the oceanic sink

Few things in science are as cut and dried as people would like them to be – or as they might seem from reading popular accounts of scientific achievements. There is always a temptation, for the

*The fact that 1976 was a very warm summer in England just shows the importance of taking *global* averages.

writer, to wrap everything up in metaphorical pretty pink bows, and to say, in effect, 'This is what researchers have discovered, that is what it means, and here's what it will do to your life.' But if it were really all so neat and tidy, there wouldn't be any science left to do at all. Science is all about investigating problems that we *don't* have the answers to, yet. And when we do find the answers to the questions we have been asking, it usually turns out that they open up new questions, which maybe nobody had ever thought of before.

I have emphasized that all climatic 'forecasts' are, in fact, scenarios – models of the way the world *might* be in the future, if certain assumptions turn out to be valid. Actually, the whole of science is like that: even our most treasured and valuable scientific theories, such as quantum physics or the cosmological theory of the Big Bang, are really self-consistent scenarios, models that describe the workings of the world in a way that matches every test we have been able to apply. But that doesn't mean that they might not, one day, be superceded by better models, if and when new discoveries are made that cannot be fitted into the old framework. Although it is more limited in scope than either of those great theories, the greenhouse effect is a very good model that undoubtedly tells us a great deal about the real world in which we live, and how it will change in the next few years and decades. But it is important to appreciate that it is not some ultimate truth, a prediction that *must* come true. Most importantly, *we* can change the forecast ourselves, if we choose to follow paths that lead to either more or less emission of greenhouse gases than the standard scenarios anticipate. But we also need to encourage more scientific research into key aspects of the problem in order to be able to make better forecasts of the way the world is going. Most especially, we need to know a lot more about the interactions between carbon dioxide in the air and in the oceans.

The first problem is that we don't really know how much dissolved organic carbon there is in the sea. In the late 1980s, two Japanese researchers, Yukio Sugimura and Yoshimi Suzuki, reported that they had developed a new technique for measuring the traces of organic material dissolved in sea water, and that this showed the presence of a whole class of large organic molecules that had never been detected by traditional techniques. These must be the remains of living organisms, stored in solution instead of being eaten by other creatures or falling to the bottom of the sea. Previous studies of the amount of DOC in sea water were largely based on a

169

technique involving the oxidation of several hundred millilitres of sea water and analysing the chemical compounds produced (carbon dioxide being among the simplest of them). The new Japanese technique uses very small samples of sea water, as little as a hundred microlitres, 'burnt' in a high temperature furnace and analysed by spectroscopy, using the characteristic 'fingerprints' of the lines in the spectrum produced by each compound. It is easier to be sure that such a small sample is oxidized completely, but much harder to be sure that all the products have been caught and analysed. When the Japanese team carried out the analysis, they found that there may be four times as much DOC in surface water (the top few hundred metres) as previously suspected, and perhaps twice as much as the traditional techniques show in deep water.

Other researchers are still trying to come to terms with this discovery. Some have sought to reproduce the Japanese measurements, with mixed success. Others are trying to find out what, if anything, is wrong with the traditional techniques. But if the measurements are taken at face value, they imply that the amount of dissolved organic carbon in the oceans is about twice the amount in the atmosphere as carbon dioxide, and twice the amount stored in biomass as plants on land. This makes it the biggest single pool of potential carbon dioxide easily accessible to the atmosphere, and if the size of that pool were to grow or shrink this could cause a significant change in the amount of carbon dioxide in the air, and therefore in the strength of the greenhouse effect.

How seriously should this new work be taken? The best independent evidence that the studies are pointing to an important new understanding of the oceans comes from computer simulations of the circulation of organic material through the oceans, carried out by Robbie Toggweiler of Princeton University. The same Japanese technique also shows much higher concentrations of dissolved organic nitrogen compounds in sea water than was previously suspected, and this in itself backs up the carbon measurements, since carbon and nitrogen are the two most essential building blocks of living molecules. Computer models incorporating the new figures for both carbon and nitrogen work much more like the real world, in terms of the distribution of nutrients and carbon dioxide in the different layers of the oceans, and the speed and efficiency of the ocean's chemical cycles, than computer models based on the old figures – the essential point is that dissolved organic material takes decades to transfer from the surface layer to the deep water, whereas particles of organic matter fall out much more rapidly. In the

170

old models, the downward flow of organic matter was dominated by particles; in the new models, dissolved organic matter is every bit as important. This, says Toggweiler, 'represents a present day wild card in our efforts to understand the ocean's chemical cycles'. And understanding those cycles is essential to an understanding of how much carbon dioxide will be taken up by the oceans in the next few decades.

One of the few teams that has yet been able to reproduce the Japanese measurements of high concentration of DOC in surface waters is John Martin and Steve Fitzwater, the same researchers who recently suggested that iron from wind-blown dust might be the missing ingredient that enables the high latitude oceans to become more productive biologically, and therefore to take more carbon dioxide out of the air, during ice ages. Biological nitrogen fixation, a key step in building up living molecules, is a reaction that requires iron. If these studies are as reliable as they seem to be, they could be telling us that the amount of dissolved organic nitrogen, as well as dissolved organic carbon in the oceans, also depends crucially on the amount of iron available. It seems that what stops a rapid increase in the amount of phytoplankton in the cold, nutrient-rich waters at high latitudes – and a corresponding decrease in the carbon dioxide content of the atmosphere – is a shortage of iron. Dumping scrap in the oceans is, however, unlikely to be the solution to the greenhouse effect. Even during an ice age, the amount of carbon dioxide 'drawn down' into the oceans was not as much as the amount human activities will be putting into the air over the next century – it was scarcely more than the amount we have already added to the atmosphere since the mid-nineteenth century – and in any case stimulating phytoplankton into more vigorous activity has no beneficial effect on other greenhouse gases such as CFCs and methane.

Researchers do not, in fact, know anywhere near as much as they would like about the biological productivity of the oceans. Recent measurements in the central north Pacific, some 640 kilometres north of Hawaii, have shown the plankton to be so active that they are taking up 550 milligrams of carbon per square metre per day; other measurements from the same area give figures only half or a third as big. Oceanographers now believe that the discrepancy may indicate that the biological activity of the oceans goes in pulses, and that many research cruises have missed these pulses of high productivity. Overall, the productivity of the oceans is probably higher than used to be thought, but not at the peak levels, measured on some cruises, every day of the year. Biological activity probably also explains another puzzling recent discovery, that the north Pacific Ocean may be able to take up *more* carbon dioxide when it is warmer (in summer) than when it is colder (in winter). Conventional wisdom has it that colder water can dissolve carbon dioxide more easily than warmer water, and this is undoubtedly true, other things being equal. It is one of the other wild cards that may cause the greenhouse effect to accelerate unpredictably as the world warms. In the north Pacific, however, other things are not equal. Simple dissolution of carbon dioxide is not the only way the gas gets taken out of the air and into the water.

Taro Takahasi, of the Lamont-Doherty Geological Observatory of Columbia University, is a member of a team that has been taking measurements of oceanic carbon dioxide throughout the year from cargo ships crossing the high latitudes of the Pacific Ocean. Prior to this study, the interaction between carbon dioxide and the ocean in winter at high latitudes was virtually unexplored, but it was expected that because carbon dioxide dissolves more easily in cold water, and cold water can be carried deep into the ocean by convection, these winter waters would be a strong sink for the gas. In fact, the tests show that most carbon dioxide is being absorbed in high summer, from July to September. In winter, the north Pacific water is actually a *source* of carbon dioxide, which is being given off by the cold, carbon dioxide rich water. The enrichment of carbon dioxide by upwelling in winter more than compensates for the increased ability of atmospheric carbon dioxide to dissolve at the lower temperatures. Two factors contribute to the ability of the ocean to absorb carbon dioxide in summer. One is the increased photosynthesis by plankton during the summer months; the other is that in summer the surface layers are warm, and so convection, which would carry deep water rich in carbon dioxide to the surface, is suppressed. Over the course of a whole year the northwestern Pacific may be a net source of carbon dioxide, releasing the equivalent of as much as five per cent of the amount of the gas produced by industrial activity.

Perhaps, then, when the world warms *some* parts of the ocean may be able to absorb *more* carbon dioxide than they do today,*

*This is actually rather unlikely, since what stimulates the plankton to grow more vigorously in summer is almost certainly the availability of more sunlight for photosynthesis, not the warmth. In a greenhouse world there will probably be more clouds, reducing the availability of sunlight and decreasing the efficiency of this particular carbon dioxide sink.

alleviating the greenhouse problem (or, at least, alleviating the less than half of the greenhouse problem that is caused by carbon dioxide). But a lot more studies from other parts of the world, especially the Southern Ocean, will be needed before anyone can be sure exactly how and where the fifty per cent of the carbon dioxide produced today that is *not* staying in the air is getting into the oceans. It must be going somewhere, and if the northwestern Pacific is a net source of carbon dioxide, sinks in other parts of the world must be more efficient than researchers had realized. They need to identify those sinks, and find out how they work, before they can make any predictions of which way their influence will change as the amount of carbon dioxide in the atmosphere builds up, and global temperatures increase.

One likely sink that needs further study is the region of the oceans that borders the land. Continental shelves make up nearly eight per cent of the area of the ocean, and they are sites of strong biological activity. Some of the carbon in organisms that live on the shelves is recycled as living things eat one another, but some escapes as biological debris over the edge of the shelf and into the deep ocean, where it may be stored for a very long time. This loss can only be balanced by a net uptake of carbon dioxide from the atmosphere, and some calculations (by John Walsh of the University of South Florida) suggest that the net 'export' of carbon in this way may be as much as a billion tonnes a year. A great deal of this activity may have been stimulated by the availability of nitrogen from human activities. These now 'fix' about half as much nitrogen from the atmosphere each year, for use in agriculture, as all the nitrogen extracted by life in the sea and on the land; a great deal of the resulting nitrogen in man-made compounds is eventually washed into the sea, and profligate use of nitrogen-based fertilizers may be responsible for about a third of the export of organic matter from the continental margins to the deep sea today.

One common thread running through all of this recent work is the importance of living things to the carbon cycles: at the same time that it is discovered that less carbon dioxide dissolves in the high Pacific through physical processes, it turns out that there is more dissolved *organic* carbon. The living components of Gaia are implicated ever more deeply in the cycles that maintain the natural greenhouse effect. But none of these uncertainties and new discoveries should induce any feeling of complacency about the ability of the world's living systems to absorb carbon dioxide and save us from our own folly. As Jim Lovelock has pointed out, the

173

simplest way for Gaia to solve the problem of human interference might be to get rid of human beings. The bottom line is still that about half of the carbon dioxide released by human activities each year stays in the atmosphere. *Whatever* the natural sources and sinks, and no matter how well or how badly we understand the interactions between carbon dioxide in the air and in the oceans, that is the basis for projections of the growth of the carbon dioxide greenhouse effect in the immediate future. And the total increase in the greenhouse warming of the globe will be roughly twice the effect produced by carbon dioxide alone over the next thirty to fifty years. It would be good to be able to resolve all the uncertainties and improve the forecasts; with any luck, the scenarios will indeed get more realistic as further research is carried out in the 1990s and beyond. But having acknowledged that the uncertainties are real, we are still left with a very solid and reliable prediction that if emissions of greenhouse gases continue at anything like present rates the world will get warmer, and quickly. And since this chapter has focused on the role of the oceans in contributing to that warming, it seems appropriate to look next at how the oceans themselves will respond to the warming. In a nutshell, when the world warms, sea level rises – but not for the reason that most people think of first.

EIGHT

The Flood Next Time

Half of the human population of the world lives in coastal regions. These are the most desirable regions of our planet, in human terms, and therefore they are already under pressure from increasing population and the problems it brings, such as pollution and destruction of the natural environment. Sea level is rising today, and has done so over the past century at a rate of about one centimetre a decade. But as the greenhouse effect accelerates, and as the warming already built into the climate system by past emissions of greenhouse gases takes a grip, a rise of at least thirty centimetres by 2030 is certain. That corresponds to a rate of increase in sea level seven times faster than over the past century; and some experts suggest that the increase could be four or five times *more* rapid, taking sea levels up by a metre and a half long before the end of the twenty-first century.

At the same time, most of the world's beaches are eroding. About seventy per cent of the beaches in the world today are being destroyed by the action of waves and weather, and very little new beach is being built up. This is partly because of the rise in sea level. For every one centimetre rise in sea level, one *metre* of sandy beach will erode, while for each ten centimetre rise in sea level the boundary between salt water and freshwater at the mouths of rivers moves one *kilometre* inland (with an associated advance of salt water into underground deposits of freshwater). Given time, if the rising trend stopped, new beaches would form at the higher level. But the rise now proceeds faster than new beaches can form. In many parts of the world human interference also contributes to the loss of beaches, and prevents them re-forming. Harbours, sea walls and the like alter the natural flow of currents and the material which builds up beaches along the shorelines; such constructions may hold back the sea as its level begins to rise, but they provide no sites where new beaches can form.

In many ways, this rise in sea level is the most tangible consequence of the anthropogenic greenhouse effect, and one which can be converted into an economic cost, in many parts of the world. London's flood protection scheme, the Thames Barrier, was recently completed at a cost of £500 million, and is designed to cope with flood tides of a severity expected, from studies of historical records, once in eighteen hundred years. But it may already be obsolescent. In the summer of 1988, a spokesman for the Thames Water Authority acknowledged that on the basis of current scientific evidence 'we shall have to produce a plan within five to ten years to heighten the barrier' if it is to cope with the high tides now expected in the middle of the next century. He didn't mention the cost of such improvements to the barrier – but how much is it worth to save the centre of London from inundation? Also in Britain, the city of Hull, for example, is already below high tide level, as is a large part of East Anglia. A rise in sea level of fifteen centimetres would double the likelihood of severe storm surges in the North Sea, bringing floods to such low lying regions. Scientists from the Institute of Terrestrial Ecology suggest that if the sea level rise anticipated over the next six decades is to be held at bay, the nation will have to spend between five and eight billion pounds on sea defences, and start work almost at once.

In the United States, an Environmental Protection Agency report, also published in 1988, took a slightly longer term view, at a scenario for projected rise in sea level of one metre by 2100. It concluded that the cumulative capital cost of protecting currently developed areas of the US would be between $73 billion and $111 billion, at 1988 prices, and that even then seven thousand square miles of dry land, an area the size of Massachusetts, would be lost beneath the waves. Losing the equivalent of Massachusetts would be a disaster even for a country as large as the US; other countries, however, stand to be wiped out entirely. The Maldives, a chain of two thousand (mostly uninhabited) atolls in the Indian Ocean, nowhere rises more than two metres above sea level today; the population of the Maldives is 177,000. In the Pacific five island states – Tokelau, the Marshall Islands, Tuvalu, the Line Islands and Kiribati – will be washed from the map.

These are not projections to be made lightly. So what is the

176

evidence that sea level has been rising in recent decades, and will rise even more rapidly in the decades ahead?

The upward trend

One of the landmark studies which picked out the upward trend in sea level over the past century was carried out by a team from the Goddard Institute for Space Studies (GISS) and published in 1982. Their interest in changes in sea level stemmed from the GISS work on the greenhouse effect – Jim Hansen was a member of the team – and compared changes in sea level in recent decades with changes in global temperatures and the expected expansion of sea water as the world warmed. Although much is sometimes made of the possibility of a massive rise in sea level as the polar icecaps melt, according to most calculations this thermal expansion of sea water is, in fact, the main reason for the rise over the past century, and will be a major factor in the decades ahead.

The techniques used to pick out the rising trend in sea level are very similar to the techniques used by Hansen and his colleagues (and others) to pick out the rising trend in temperatures. The data comes, in this case, from tide-gauge measurements made at coastal stations around the world. Like the temperature data, it shows many variations, and the records from some sites have had to be discarded because they only go back for a couple of decades, or because they come from geologically active regions such as the Pacific coast of Japan. That leaves just 193 stations, scattered around the world, with reliable long-term records that can be analysed by well established and reliable statistical techniques.

The 'raw' global sea level trend that emerges from this analysis is a rise of twelve centimetres between 1880 and 1980. But the GISS team did not take this at face value. They knew that geological evidence shows that there has been a long, slow rise in sea level since the warm period, six thousand years ago, that followed the latest ice age. This may be associated with a decline in the volume of the Greenland and Antarctic ice sheets, or it may be because the continents are still recovering from the weight of ice that used to press upon them. During an ice age, regions of the Earth's crust covered by ice sheets sink downwards, under the weight, pressing into the semi-liquid layers below the crust. Because the crust itself is solid, as one part of a continent is pushed downward in this way, the opposite end, where there is no ice, is levered upward. When the weight of ice was removed at the end of the latest ice age, regions such as Scandinavia began to lift upward, released from the burden, while regions further south began to sink downwards in a seesaw effect that continues to this day. Scotland, for example, is still rising, while southeastern England is sinking as the whole of Britain pivots about a roughly northeast/southwest line (running through the Isle of Man), adding to London's flood problems. Whatever the exact cause of the long-term trend in sea level, it can be picked out from the geological record, and its contribution to the rise over the past hundred years subtracted from the tide-gauge data. When they did this, the GISS team was left with a rise of ten centimetres per century, just one millimetre a year over and above the long-term trend.

Sea level rose over the past century in every geographical region studied by the GISS team except Scandinavia, where the recovery from the weight of ice is proceeding most strongly today, and, perhaps, the west coast of South America, which is itself a region of geological activity where the Andes are still being thrust upward, so that it is hard to distinguish any sea level rise. It is a genuine global trend. When the team compared the sea level trend with the temperature trend determined by Hansen and his colleagues (see *Figure 1.5*, Chapter One), they found that the two sets of figures showed the same pattern of behaviour, but with the sea level rise lagging about eighteen years behind the temperature rise, exactly as expected in view of the thermal inertia of the surface layer of the ocean. The GISS team calculated that half of this rise can be explained in terms of the thermal expansion of the oceans, suggesting that the rest must be due to the slow melting of glaciers at high latitudes and on mountains in low latitudes. Sea level, they concluded, is now 'at or near its highest level since the previous interglacial, the Eemian 120,000 years ago'. And they projected that by 2050 thermal expansion alone would cause a further rise in sea level of twenty or thirty centimetres (assuming that a doubling of carbon dioxide produces a rise in mean temperatures of 3°C). If this future thermal expansion is also accompanied by slow melting of ice sheets in the way that the rise over the past century has been, the rise in sea level by the middle of the next century will be between forty and sixty centimetres.

But this is just one study among many. The SCOPE 29 report compared all of the available studies of past sea level rises and came to some more broadly based, though only slightly different, conclusions. That report also looked to the future – not only on the basis of extrapolating past trends and patterns of oceanic behaviour, but on the much more difficult basis of calculating how the great ice sheets are likely to respond to the increase in global temperatures. The results are surprising.

Scope for change

The simplest assumption on which forecasts of future rises in sea level may be based is that there will be the same relationship with temperature increases as in the past. If a warming of half a degree Celsius has produced a rise in sea level of ten centimetres, then the next warming by half a degree ought, on this picture, to do the same. And *perhaps* it isn't too wild a guess to suggest that a warming of 3°C will cause a rise in sea level six times as big, by sixty centimetres. This approach is certainly justified when thermal expansion of sea water is the main driving force behind the sea level rise, but it becomes less sound the more other factors are brought into play. The greatest weakness of this approach comes when we use it to extrapolate from a change caused by a warming of only a fraction of a degree in a hundred years to changes caused by a global warming of several degrees in a matter of three or four decades. Even so, it does give a first guide to the size of the problem we face.

The GISS study, it turns out, gives estimates at the low end of the range considered in the SCOPE report. Other similar studies give estimates of the rise in sea level between 1880 and 1980 as high as fifteen centimetres – but they do not all take account of the long-term trends that the GISS group subtracts from the data. More detailed analysis of the trend in sea level shows a rapid rise from about 1920 to 1950, a levelling off from about 1950 to 1970 (following the break in the warming trend in the mid-twentieth century), and a renewed rapid rise since 1970. But overall the rise seems to be speeding up. Analyses of sea level variations that are based only on data from 1930 onwards, and ignore the first half of the period studied by the GISS team, show a much more rapid increase than they found, as high as the equivalent of twenty to thirty centimetres a century.

Because of this spread of estimates of how rapidly sea level is responding to the present rise in temperatures, even a straight-forward extrapolation of the trend into the twenty-first century gives a corresponding range of forecasts. Allowing for the eighteen-year lag in the response of the oceans to rising temperatures, the GISS study implies an increase of sixteen centimetres in sea level for every degree Celsius rise in temperature, while the highest figure accepted in the SCOPE study corresponds to an increase almost twice as rapid, thirty centimetres per degree. Using the estimate that a doubling of the carbon dioxide concentration will increase global mean temperatures by 3.5°C, the SCOPE report suggests that this will produce a rise in sea level of between fifty-six and 105 centimetres, but that the uncertainties involved in all the calculations are so large that any range from twenty-five centimetres to 165 centimetres cannot be ruled out. As we saw in Chapter Six, the equivalent of a doubling of the natural carbon dioxide greenhouse effect is likely to happen in the decade of the 2020s, if no action is taken to slow the increase. This is the basis of the forecast that the *minimum* rise in sea level by 2030 will be thirty centimetres; the *most likely* rise is about twice that figure.

But this forecast is based on the assumption that the straightforward relationship between temperature and sea level will hold, and in particular on the assumption that most of the future contribution to the rise in sea level will continue to come from thermal expansion of the oceans, which does indeed respond in this 'linear' way to increasing temperatures. This is certainly a big effect, but how big? For a rough guide to what is going on, oceanographers use observations of the region of the ocean between 45°N and 45°S, which contains most of the water of the world. The surface layer of the ocean, which is well mixed up and therefore effectively all in contact with the air above, is about a hundred metres deep; a 3°C rise in temperature of this layer would increase its thickness by around nine centimetres, and the half a degree warming from 1880 to 1980 should have thickened it by just one and a half centimetres.

The next layer of the ocean extends down to about a thousand metres, and is formed from cold water at higher latitudes that slides underneath the warm surface layer. The surface layer is heated directly through its contact with the atmosphere and by the direct rays of the Sun, but the deeper layer is imprinted with the temperature at high latitudes, where it formed, *at the time it formed*. In the Atlantic between Bermuda and Spain, this layer cooled by about half a degree Celsius between 1957 and 1981, probably responding to the cooling of the northern hemisphere between 1940 and 1960. That may have thinned the layer between a hundred and a thousand metres by about four centimetres. Even deeper water, fed even more slowly by water from the highest latitudes, in the Arctic, warmed by about 0.2°C over the same period, probably responding, sluggishly, to the arrival of water that participated in the warming of the Arctic *before* about 1940. That warming would have caused the top two thousand metres of deep water to expand by about four centimetres, almost exactly cancelling the thinning of the layer above. Even deeper water has probably not been affected by any of these relatively rapid fluctuations in surface temperature, but will warm and expand very slowly if the excess greenhouse effect persists for centuries and millennia. Just by chance, the overall increase due to thermal expansion between 1958 and 1981 should have been roughly equivalent to the contribution from the surface layer alone. Unfortunately, data from 1957 and before, which might have helped to test these ideas, simply does not exist.

These complications highlight the difficulty of making forecasts of future trends. Researchers have to assess the future run of temperature changes on different timescales, in different layers of the ocean, and add them together in some appropriate fashion. But on the assumption that there will be no change in the structure of the ocean and its density distribution, Roger Revelle estimates that a global warming of 4.5°C over the next hundred years would increase sea level by at least thirty centimetres due to the expansion of water alone.

At this point, it is only fair to say that there is some doubt about how much thermal expansion may have already contributed to the rise in sea level, and therefore how strong the effect will be in the twenty-first century. One respected researcher disagrees with the forecast sea level rise projected on this basis by the SCOPE and GISS (and other) teams. Tom Wigley, of the University of East Anglia, reported in 1987 the results of calculations he had carried out with Sarah Raper. They imply a contribution from thermal expansion alone of only a four to eight centimetres rise in sea level by 2025. There are several reasons for this apparent discrepancy. First, the team has used a different technique to calculate how much thermal expansion has already contributed to the rise in sea level from 1880 to 1980 (in fact, to 1985). Wigley says that these calculations are more sophisticated, and more reliable, than those used in previous studies, because they do not treat the oceans as distinctly separate layers but include a proper allowance (or at least, *some* allowance) for upwelling from the deep ocean. The contribution of thermal expansion to the rise between

1880 and 1985 comes out as only two to five centimetres in these calculations, less than half the total actually recorded. This in itself reduces the size of the amount of rise due to this effect forecast for 2025, which comes out still lower because Wigley and Raper assumed a temperature rise of no more than 1°C, very much at (or below) the bottom end of current estimates. All this leads to their forecast of a contribution from thermal expansion of a rise in sea level of four to eight centimetres between 1980 and 2025. The equivalent GISS forecast (for the contribution from thermal expansion, not the total rise in sea level by 2025) suggests a figure of ten centimetres, and is itself lower than some other estimates.

The disagreement is an example of science at work, with different researchers using different techniques to try to peer into the future. Stephen Schneider has said that the process is like trying to gaze into a dirty crystal ball. By taking time to clean the glass (that is, improving the scientific models) you can get a better picture; but at some point it is necessary to decide that the picture is good enough to alert policy makers and the general public to the hazards ahead, even though they can still only be seen imperfectly. That point has certainly been reached with studies of the greenhouse effect in general and the prospect of rising sea levels in particular. For, remember, the suggestion that thermal expansion may have contributed less than we thought to the *measured* rise in sea level in recent decades only means that melting of glaciers and icecaps must be contributing more. There is no doubt that the sea level is rising, but if thermal expansion is playing a lesser role, then forecasts based on straightforward extrapolation of the past trend become less reliable. If melting of ice is becoming the dominant factor, then it is likely that the increase in sea level will proceed more rapidly than the straightforward extrapolations imply, since, as we have seen, there is a positive feedback at the edge of ice sheets whereby melting of the ice uncovers more dark land, which absorbs more solar heat and melts more ice. Estimating future changes in sea level depends crucially on estimating changes in the ice cover of the globe, and though the crystal ball may be dirty any information we can glean from it is invaluable.

Icemelt

After the oceans and lakes, ice provides the biggest reservoir of water on our planet today. The amount of water vapour in the air at present, on average, is equivalent to a layer of water just twenty-five millimetres thick over the entire surface of the globe, or thirty-five millimetres over the oceans alone. If the temperature at the equator increased by 3°C, while the temperature increase at high latitudes was twice as big, the amount of extra water evaporated from the oceans would cause a fall in sea level of just seven millimetres. Man-made reservoirs contain even more water, the equivalent of a layer of ocean 1.5 centimetres thick, but these are semi-permanent features that do not affect the long-term balance of water in the seas. About two-thirds as much water is taken out of river systems and used in irrigation each year; but, of course, this water soon runs back into the sea, or evaporates (or is breathed out by plants) and joins the moisture in the air. New reservoirs are still being built, and stores of ground water are being extracted by human activities, but these are not major influences on sea level; according to the SCOPE report, on balance human management of water stocks is causing a decline in sea level of about one centimetre per century. Like the change caused by the increase in the moisture content of the air, this can be ignored compared with the changes in sea level caused by thermal expansion and by icemelt.

Glaciers and ice sheets cannot be ignored. They store about three-quarters of all the fresh water in the world, and cover eleven per cent of the land area of the globe. If all this ice melted, it would raise sea level by over seventy metres. Ice sheets, and their responses to rising temperatures, are not as well understood as we would like. The experts still debate key ideas among themselves, and disagree on many issues. Most of these disagreements are not important in the context of the present book, and rather than trying to give every theory a fair crack of the whip – which would mean writing another book at least as long as this one – I shall focus on the interpretation of the evidence provided by one of the most respected authorities, Gordon Robin, of the Scott Polar Research Institute in Cambridge.*

Greenland and Antarctica contain, between them, about 99.5 per cent of all the land ice on our planet today. But in the short term the smaller glaciers and icecaps in other parts of the world contribute a disproportionate amount to the run-off of water into the sea, because they are at warmer latitudes and are more susceptible to icemelt. Studies of the way in which these smaller

*Robin is the main contributor on the subject to the SCOPE 29 study, which is the principal source for this section.

glaciers have responded to temperature variations in the twentieth century suggest that when the world warms by a further 3.5°C enough ice will melt to raise sea level by about twenty centimetres. This figure is probably an overestimate, because changes during the present century have occurred mostly in the smaller glaciers in temperate latitudes, and larger, colder glaciers will resist the warming trend more effectively. There is also likely to be a considerable delay in the melting back of glaciers – if temperatures rise by 3.5°C by 2030, as they well might, the glaciers will not come into equilibrium with those temperatures for decades, by which time the greenhouse effect will have made the world warmer still. In any case, though, melting *all* of the glaciers and icecaps outside Greenland and Antarctica would only raise the sea level worldwide by thirty-three centimetres. This would happen, given enough time for the warming to penetrate to the hearts of the biggest of these glaciers, for a rise in global mean temperatures of just 6.5°C, which could occur before the end of the twenty-first century.

Ice floating in the sea does not come into these calculations at all, because when this ice melts the volume of water it produces is almost the same as the volume it displaces when it is in the form of floating ice, in exactly the same way that the level of liquid in a tumbler of water laced with ice cubes does not change when the ice cubes melt (this is, essentially, the discovery that made Archimedes so excited that he leaped out of his bath with a cry of 'Eureka' and ran naked through the streets to spread the news). So that leaves Greenland and Antarctica as the sources of melting ice that could have the biggest impact on sea level. The east Antarctic ice sheet contains the equivalent of a fifty-five metre rise in sea level, while melting all of the ice in west Antarctica would raise sea level by six metres, and melting the ice over Greenland would produce an increase of eight metres. To put these figures in perspective, a rise in sea level of eight metres would be just enough to enable the President of the United States to launch a boat from the south lawn of the White House and row along the coast to the steps of the US Capitol building to visit the Senate. But don't worry too much – this is *not* going to happen in our lifetimes! What is going to happen depends on a combination of factors, which work in similar ways but may produce different results in Greenland and in Antarctica.

These ice sheets are not just great lumps of ice sitting on the land beneath, but are dynamic features of our planet. Inland, buried beneath the ice, there are great mountain ranges, and high altitude plateaus, where the air is even colder than it is at the edge of the ice. There snow falls and accumulates in winter, building up a mass of ice, over thousands of years, that slides slowly out of the hinterland and downhill to the sea in the form of glaciers – rivers of ice flowing at speeds ranging from a few centimetres to a few tens of metres a year. In the interior of an ice sheet, where the ice is thickest and the downward slope is smallest, the flow is very slow; as it moves outward, the rate increases to a few tens of metres a year, and even to several hundred metres a year, as the slope steepens and the ice thins (the same *volume* of ice is flowing out at each distance from the centre, even though in some places this is a thick sheet moving a few centimetres and in others it is a thin sheet moving, say, a hundred metres; the ice gets thin because it stretches under its own weight as it flows downhill). At the edge of the Ross Ice Shelf, in Antarctica, ice is moving out into the ocean, where it breaks off and floats away, at a rate of nine hundred metres a year. But it takes thousands of years for snow that falls in the centre of Antarctica to escape from the continent as an iceberg, and because the temperature of the world is always changing, the ice sheets are never really in equilibrium with the temperatures over a century or so. The icecaps today still have a partial memory of the latest ice age; during the Last Glacial Maximum, there was ice in Antarctica that had formed during warmer times.

Three factors together determine the way in which global warming will cause a major ice sheet to make a contribution to the changes in sea level. First, when the world is warmer and there is more water vapour in the air, there is more snowfall over the ice sheets (as long as the world is still cool enough for snow to fall at all). The great ice sheets are likely to get *thicker* as the world warms. Secondly, some ice melts at the edge of the ice sheets. So the area covered by the ice sheet is likely to *decrease*. Finally, the rate at which glaciers flow out from the ice sheets and break up will change. Because floating ice displaces its own volume of water, a lump of glacier that moves off the land and into the sea is effectively the same, as far as sea level is concerned, as the same volume of water added to the ocean. The flow of glaciers will be altered as the ice sheets become smaller and fatter. By and large, ice will flow more rapidly out from the central, higher regions of the icecap; but calculations have to be carried out carefully in each case, allowing for the topography of the underlying land, before it is possible to say just how this will affect sea level in the short term.

The Greenland icecap has been getting thicker in the middle and thinner at the edges in recent decades, which broadly fits the expected pattern for a warming world. But there are too few reliable measurements to be sure what effect, if any, this has had on sea level. The 'best guess' among the experts seems to be that the two effects more or less cancel each other out, but that Greenland's contribution to changes in sea level at present may be equivalent to either a rise or a fall, at a rate of no more than three centimetres per century. Change in Antarctica can have a much bigger effect because there is so much more ice there – ice based on the *land* surface of the Antarctic continent (leaving aside the sea ice) covers an area one-thirtieth as big as the area of all the world's oceans. But as in Greenland, what matters is the balance between the rate at which new ice is forming from snow in the interior and the rate at which old ice is flowing off the land at the edges of the icecap. Once again, it seems that the two effects are almost in balance today, but some evidence, from satellite measurements, suggests that the outflow is not quite keeping pace with accumulation, and that the contribution of Antarctica to sea level changes today is a small *decrease*, probably a few centimetres per century.* Ice core studies show that the rate at which ice was accumulating each year during the Last Glacial Maximum was only between a quarter and half of the present rate, and Robin concludes that a global warming of 3.5°C should *increase* the rate of accumulation of ice in Antarctica by at least ten per cent and probably by more than twenty-five per cent above the present level. Melting of large quantities of ice at the edge of the icecaps in Greenland and Antarctica may well occur in our lifetimes, but the effects on sea level will be roughly balanced by the accumulation of new ice in their interiors.

These estimates are borne out by more detailed studies that calculate the changes in ice flow caused by, in effect, taking mass away from the edge of the ice sheet at sea level and putting it in the middle of the ice sheet at high altitude. In Greenland, if the average temperature of the world were increased by 3.5°C and

*If the temperature stayed the same long enough for the ice sheets to come into equilibrium, then the accumulation in the centre would exactly balance the outflow. If more snow accumulates, there is a greater weight of ice pushing down to the sea, which speeds the glaciers up. But this process takes thousands of years to reach a new balance; on the timescales that matter to us, up to the end of the twenty-first century, there is no time for this effect to have much influence.

then held steady at that new value, the immediate overall effect would be to dump 110 cubic kilometres *more* ice into the sea each year, raising sea level by about three centimetres in a century. In Antarctica, the balance works the other way, with about a hundred cubic kilometres of extra ice being added to the polar cap each year, corresponding to a *fall* in global sea level of 2.8 centimetres in a hundred years. Taking both polar regions into account, the net effect is small, and may cancel out entirely; allowing for uncertainties in the calculations, the main ice sheets of the world will between them produce either a rise of a few centimetres or a fall of a few centimetres in sea level over the next century. After allowing for thermal expansion of ocean water, most of the rise in sea level over the past hundred years has come from the melting of small glaciers at lower latitudes, and this will still be the case throughout the twenty-first century, and until all of the ice locked up in these small glaciers has gone. Rising sea levels today are not caused by melting of the great polar ice sheets, and if anything the polar ice sheets are helping to reduce the rise in sea level. So why do stories of dramatic flooding caused by a break-up of the Antarctic ice still sometimes make headlines? Apart from the fact that such headlines help to sell newspapers, it is because there is a possibility, according to some researchers, that the small, steady changes I have been describing so far might be overwhelmed by a catastrophe, a disaster caused by a change in the way ice flows out of west Antarctica in particular. The argument about this still rages, but the weight of evidence, from both glacier theory and measurements, now seems to be that this is not on the agenda for the twenty-first century. Since the story still causes concern, however, it seems worth taking time out to try to knock it on the head.

Catastrophe not on the agenda

In the very long term, if the temperature of the world rises by about 10°C and stays there, then eventually all the ice sheets on Earth today will melt, even with the present day geography and pattern of ocean currents. But 'eventually' means several hundred years, perhaps several thousand years, even for such a large increase in temperature. So climatologists and oceanographers were puzzled when they found evidence that during the previous interglacial, 125,000 years ago, sea levels seem to have been about seven metres higher than today in equatorial regions of the globe. According to some ice core studies, the temperature then was about 3°C higher than today, but that shouldn't have been enough to cause such a large sea level rise; and, in any case, other geological evidence suggests that the climate of the previous interglacial was very similar to the climate of the present one. This is now a solid conclusion based on analyses of isotope ratios from fifty-two sea bed cores drilled at sites around the world. So what did happen to sea level in the previous interglacial?

The catastrophists largely ignore the data from the fifty-two sea bed cores, and take at face value the ice core data that seem to imply a 3°C higher temperature. Even then, they have to use considerable ingenuity in order to get a global flood of such proportions. The idea rests upon the way in which the west Antarctic ice sheet itself rests on a series of islands where it stretches out into the ocean. These islands support the bottom of what would otherwise be sea ice, like the fingers of your upturned hand supporting a plate, or like the pillars supporting the roof of a cathedral. Because the ice is grounded in this way, there is more friction holding back the weight of ice pressing down from the heights of Antarctica than there would be if it were floating on the sea. The catastrophe scenario suggests that as the world warms and sea level rises this layer of ice might be lifted up from the pillars that support it, allowing water to creep underneath and act as a lubricant, reducing the friction. If this happened, the ice might 'surge' forward dramatically, under the pressure from above. This obviously could happen, but nobody can say how much the sea level would have to rise first in order to do the trick, nor what temperature increase that would correspond to. It is pure guesswork to argue that a supposed rise of 3°C (compared with today) would have sufficed 125,000 years ago.

More guesswork comes into speculations about what might happen if the ice did surge. One likely effect is that large amounts of ice would drift to lower latitudes, eventually melting. But the sea level would rise straight away, as soon as the ice gets into the sea, not when it melts. So there could be a period of higher sea level but with a *greater* expanse of ice over the Southern Ocean. The effect of all that ice would be to reflect away incoming solar heat, cooling the globe and, perhaps, plunging it back into a new ice age. If so, the world should have been colder just after the time that sea level rose, not warmer. These are wonderful scenarios for anyone with a taste (like me) for science fiction. But Robin argues that they have no foundation in science fact. He can explain the rise in sea level in the previous interglacial, *and* the increased temperatures recorded in the Vostok core, in another way.

Robin's version of events goes back to the ice age before last. This lasted from two hundred thousand to 135 thousand years ago, at its most intense, and was both colder and had a more prolonged cold period than the most recent ice age, which was at its height between seventy thousand and fifteen thousand years ago. I have already mentioned that it takes a very long time for the ice sheets to come into equilibrium with the prevailing temperatures. Antarctica, for example, is still trying to adjust to the warmth of the present interglacial, let alone any extra warmth contributed by human activities. Ice is accumulating at a rate of one centimetre a year, even after allowing for the outflow from the central plateau, and has done since the latest ice age ended, raising the height of the polar ice above sea level by 180 metres since the end of the Last Glacial Maximum, eighteen thousand years ago. This is only about ten per cent of the thickness of the Antarctic ice sheet, which ranges from 1,720 to 2,200 metres in thickness today. Because of the thickness of this layer, the 'surface' level of Antarctica, the *average* altitude where it is possible to stand on solid ice, is 2.2 kilometres above sea level; no other continent has an average surface level as much as one kilometre above sea level. The thickness of the layer, however, varies as the environment changes. During an ice age, when there is less snowfall, the outflow from the plateau more than compensates for the new snow falling each year, and the ice sheet thins. The longer the ice age, the more it thins; it would only reach equilibrium at a new, thinner level after well over a hundred thousand years. During the ice age before the last, which was colder and longer than the most recent ice age, it must have thinned more than it did during the latest ice age. So the thickness of ice at the beginning of the previous interglacial was less than the thickness at the start of the present interglacial, and 125,000 years ago the ice cover over Antarctica was probably at least two hundred metres thinner than the ice cover there today.

Such a difference in ice cover would mean two things. First, if there is that much less water locked up as ice in Antarctica there must be the equivalent amount of 'extra' water in the sea – just enough, in fact, to raise sea level by seven metres. Secondly, if the altitude of the ice surface in central Antarctica was two hundred metres lower than today, the temperature of the air blowing over that surface would be higher, by two or three degrees Celsius.

All of the puzzling difference in temperature can be explained if the ice surface was three metres lower 125,000 years ago than it is today (but that is not necessary if temperatures were indeed a degree or so higher than today, as other evidence suggests). So Robin concludes that there is no compelling evidence that a surge of the west Antarctic ice sheet occurred in the previous interglacial, and no reason to expect that a rise in temperatures of 3°C today would make it surge now.

These conclusions fit the geological evidence that the west Antarctic ice sheet was even bigger (that is, it contained more ice, although it may not have occupied a larger area) around four to six million years ago, when the world as a whole was warmer than it is today. They also match up with the results of computer studies of how ice flows in Antarctica, carried out for the US Department of Energy and reported in 1987. These computer simulations, which have been very successful in modelling surges of mountain glaciers at lower latitudes, show that the Antarctic ice sheet is relatively stable. Even when the researchers increased the amount of snowfall in their computer models, and reduced the friction between the ice sheets and the underlying rock, the Antarctic ice sheets showed no tendency to surge, except in two minor cases. Instead, the thickness of the ice sheets and the position of the boundary between sea and ice changed slowly and steadily when external factors were changed to match the predicted climatic consequences of a build-up of greenhouse gases in the air. One reason for this is that around Antarctica the ice streams are channelled into confined bays, which resist the outward push of the ice. Squeezing ice out of central Antarctica into the open sea is about as 'easy' as squeezing the last bit of toothpaste from a tube – it doesn't get out unless there is a strong push. That DoE study concluded that 'Antarctic ice surges do not constitute a clear and present danger', even in the light of the anthropogenic greenhouse effect.*

And yet, we know that lesser glaciers do surge, sometimes – the DoE study used data from observations of such surges. In recent years, the most publicized example was the surge of the Hubbard glacier in southern Alaska, which in May 1986 rushed forward, advancing by two kilometres in a few weeks and sealing off the sea end of a fjord fifty-five kilometres long. Until the sea

*'On the Surging Potential of Polar Ice Streams', DOE/ER/60197-H1, National Technical Information Service, Springfield, Va.

190

broke through the barrier, there was great public concern for the marine wildlife trapped in the fjord. Almost certainly, the main reason why the Hubbard glacier surged when it did was because of the greenhouse effect, producing more water vapour and more snowfall, making the glacier thicker and heavier as the decades passed, until at some critical weight it slid downhill in the glacial equivalent of a hurry. A contributory factor may have been an accumulation of water collecting beneath the ice (because of increased melting in a warmer world) and acting as a lubricant. The warm El Niño winters of the mid-1980s could have played a part in both these processes. Glaciers *do* surge when conditions change enough. What will happen if, and when, conditions change enough to destabilize the Greenland or Antarctic glaciers?

According to Robin, the Antarctic ice sheet will not be de-stabilized even by a rise in temperature of 6°C. It has been around for millions of years, including times when the world was a lot warmer than it is today, and we can rely on it still being there at the end of the twenty-first century – as far ahead as anyone can reasonably expect to look in a world changing as rapidly as ours. Greenland, however, is another matter. For a rise in temperatures of around five or six degrees, surface melting and the flow of ice into the sea will begin to exceed the rate at which new ice is forming from snow falling on the icecap, and the ice sheet will rapidly begin to diminish. The same temperature rise will also melt the ice floating on the Arctic Ocean, over the North Pole itself; this will have no effect on sea level, but will bring dramatic climatic changes, especially in the northern hemisphere, which will be discussed in Chapter Nine. Greenland itself would then be losing more than six hundred cubic kilometres of ice every year. But this 'rapid' destruction of the Greenland ice, as drastic as an ice surge, would cause a rise in sea level at a rate of only twenty centimetres a century. Even if the west Antarctic ice sheet surged, says Robin, this would still take two hundred years to raise sea level by another five metres – dramatic speed on a geological timescale, but hardly an overnight flood of Biblical proportions.

The comforting thing about the worst scenarios for a rising sea level is that these are such long, slow processes that you can walk inland faster than the water can advance behind you. The bad news is that even such insidious effects can have disastrous consequences for coastal regions, and for people who cannot, for one reason or another, simply pack up their possessions and move to higher ground. All things considered, sea level is still likely to rise by a metre or more by the middle of the twenty-first century.

Counting the cost

It is impossible to do more here than sketch in some of the consequences of rising sea level in the twenty-first century, and to hint at the economic costs involved in a few cases; but even the sketchiest overview gives some idea of the rapidity of the likely impact on human activities. The effects will be felt most severely in regions where the land is already sinking as a result of geological activity, and these include the Nile Delta, where the city of Alexandria is barely above sea level today, much of the Atlantic and Gulf coasts of the United States, and southeastern parts of England, especially East Anglia. The beautiful city of Venice might as well be written off now – it is quite simply too late to save it from the floods, and money spent trying to hold back the waves is money wasted; the only hope of preserving some trace of its beauty is to uproot the art treasures and move them away.

Since I live in the southeast of England, perhaps I can be excused for mentioning some of the problems the region faces in a little more detail. Here, the combination of icemelt and landsink may bring a total rise in sea level of several metres over the next hundred years. Not far along the coast from where I am writing, this will flood the shallow gravel ridge at Dungeness, shifting the shingle and threatening the nuclear power stations which now sit just above sea level. Virtually all of the Thames estuary will be severely affected; the inland waters of the Norfolk Broads will be

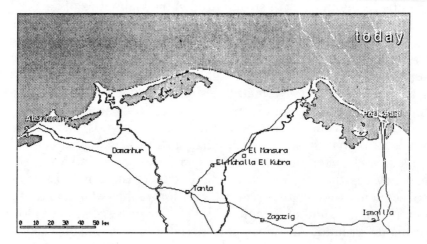

192

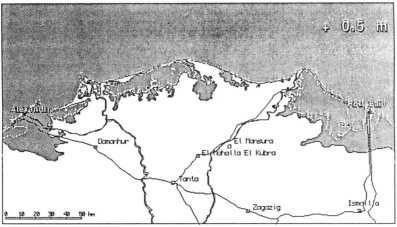

Cropland inundated: 1754 km2
Population affected: 3.3 million

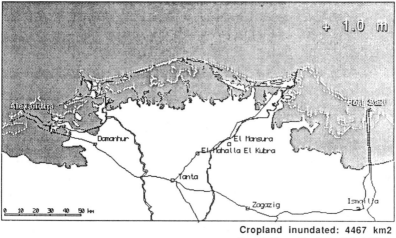

Cropland inundated: 4467 km2
Population affected: 5.3 million

Figure 8.1
These projections by UNEP show the impact of a rise in sea level of half a metre or one metre on the densely populated, low-lying region of the Nile Delta. At projected rates of global warming, the one-metre mark will be passed soon after the middle of the twenty-first century.

penetrated by salt water, destroying a valuable freshwater habitat; and reclaimed land in the Wash will be flooded as the sea outflanks existing flood defences.

Greater human tragedies are likely to occur in poor countries with large populations living near sea level – the Nile Delta comes into this category, as does the Ganges Delta in Bangladesh. A rise

193

in sea level of just fifty centimetres will flood only 0.4 per cent of the land area of Egypt, but sixteen per cent of the country's population lives in that coastal strip. A one-metre rise, quite possible by the late twenty-first century, would take out twelve per cent of the land area of Bangladesh, nine per cent of its population and eleven per cent of its agricultural land. The region will suffer doubly, because in a hotter world with more evaporation from the oceans hurricanes will be more common and more severe. As for the rest of the world, if you live by the sea, it is a sobering exercise to get hold of a contour map yourself, showing the local region, and colour in everything below one metre above sea level. You are likely to live to see that land disappear under water. But that isn't the only problem.

When the sea level rises because of a global warming, extreme climate-related events that cause flooding – hurricanes in the tropics, storms at higher latitudes – become disproportionately more extreme. The Dutch have spent the equivalent of \$2.5 billion on a massive project to protect them from storms in the North Sea, a barrier on the estuary of the Eastern Scheldt River that can be closed if catastrophe threatens. Their project, completed in the 1980s, dates back to the storm surge of 1953 which claimed 1,835 lives and flooded 150,000 hectares. Farm-lands covered in salty sand and silt took years to recover, and the economic cost has never been fully quantified. The new barrier was designed to protect the region from the kind of storm that 'ought' to occur once in a hundred thousand years; if sea level rises by one metre, *without* any increase in severity of storms, the protection will be against a once in five hundred years disaster – in other words, there will be a one in five hundred chance that the barrier might be breached in any single year. Not long odds at all, in the context, and a two-metre rise in sea level reduces them to one in a hundred.

Just as cities have often been built on or near coasts and rivers, for ease of communications, many industrial activities, including nuclear power plants, have been placed there, either to give easy access to ports or to make use of cooling sea water. Relatively modest increases in sea level hit these important economic sectors disproportionately hard. In some parts of the world, especially the US, estimates have been made of the specific economic cost to certain regions of a particular rise in sea level. If the sea rises by sixty-four centimetres by 2025, for example, the cost to Charleston, in South Carolina, would be just over a billion dollars in 1988 money, according to Environmental Protection Agency figures. Most of this would be the result of increased storm damage, like the North Sea storms that could break through the Eastern Scheldt barrier and the Thames barrier. The same sea level rise over the same time period would cost Galveston, Texas, $360 million. Sea walls, and comparable 'hard' defences – barriers against the sea – can themselves cost millions of dollars a kilometre to build, and once built they have to be maintained. It may be cheaper, in the long run, to beat a strategic retreat. But a lot depends on how you calculate the cost.

In 1980, Stephen Schneider and Robert Chen tried to put numbers into such debate, using figures which are now rather old but which still make their point. Deliberately picking figures at the high end of the range of estimates of sea level changes projected for the twenty-first century, they looked first at the impact of a five-metre rise on the US. A quarter of Florida, including all but four of its cities with populations of over twenty-five thousand in 1970, would disappear from the map. To the north, major cities such as New York, Atlantic City and Boston (including most of the Massachusetts Institute of Technology and Harvard University) would be inundated; while around the Gulf New Orleans would disappear and large parts of the Texas coast would be flooded. Using figures from the 1970 census, Schneider and Chen calculated that eleven million people (six per cent of the population of the continental US) and property valued at $110 billion (all figures in 1970 dollars) would be affected. For a rise in sea level of 7.5 metres, likely to happen in 150 years if the greenhouse effect is not brought under control, the cost would be a quarter of a trillion dollars ($2.5 x 10^{11}$). For the sake of argument, the team multiplied this figure by ten to give a global cost of such a rise in sea level of $2.5 x 10^{12}$, and then asked what it would be 'worth' to us, today, to prevent this catastrophe occurring on a timescale of a century and a half.

Investing a dollar at seven per cent interest doubles its notional value in just ten years, and would make it worth thirty thousand dollars in 150 years' time (ignoring inflation). The owner of a building valued at three hundred thousand dollars that is threatened by the kind of sea level rise expected in 150 years' time could 'insure' that his or her descendants did not suffer financially as a consequence simply by investing ten dollars today at seven per cent compound interest! So the cost of an inundation that will bring $2.5 x 10^{12}$ worth of economic damage in 150 years' time can be offset by investing just seventy-five million dollars today – the cost of a

195

few fossil fuel fired power plants. In harsh economic terms, it is 'not rational' to spend more than this to counter the effects of global flooding, even if the cost of that flooding will be $2.5 x 10^{12} by the middle of the twenty-second century. But if the same inundation were likely to occur in only twenty years, it would be worth spending $6 x 10^{11} now to prevent it happening.

Schneider and Chen use these examples to show why 'discounting' the future makes it unlikely that the present generation will invest heavily to provide a hedge against greenhouse effect losses in the future simply on economic grounds. In global terms, the problem is further complicated because there are many countries that are unlikely to be drastically affected by a rising sea level but are major users of fossil fuel. The economic costs of the rise in sea level caused by the greenhouse effect are great, but they hit some places, such as Bangladesh, Holland and Louisiana, disproportionately hard. Partial adaptation to the rise in sea level over the next fifty years or so would cost the US tens of billions of dollars, but spread over a planning and construction time of twenty to forty years. There would then be a need for more expenditure to respond to the continuing rise in sea level in the second half of the twenty-first century. Strictly speaking, this is not 'cost effective'. Schneider and Chen point out, however, that the cost to the US economy of the oil price hike in 1973 was about fifteen to twenty billion dollars a year for several years. So the country can, in another sense, undoubtedly afford to prepare for the rise in sea level. But the decision is not likely to be taken on purely economic grounds.

Although they may be the most easily calculated consequences of the greenhouse effect, the harsh truth is that economic impacts of rising sea level alone are not severe enough to justify major efforts to reduce our emissions of greenhouse gases. But the equally visible human impacts certainly provide a powerful, and some might say conclusive, *emotional* argument. The real cost of rising sea level is human, not economic. Indeed, in her landmark speech to the Royal Society in London on 27 September 1988, Margaret Thatcher said that the importance of the greenhouse effect 'was brought home to me at the Commonwealth Conference in Vancouver last year, when the President of the Maldive Islands reminded us that the highest part of the Maldives is only six feet above sea level'. In global economic terms, the 177,000 population of the Maldives count for nothing against the figures bandied about by Schneider and Chen. But their probable fate has been instrumental in moving even the

196

Iron Lady of British politics to reconsider her government's position on 'green' issues. It may not be a moment too soon, for even if you don't live by the sea, life in the hothouse of the twenty-first century now seems certain to be very different from anything you have been used to.

NINE

Hothouse Earth

Most of the problems of global environment pollution that confront us are given doom-laden, emotive names. The very terms 'acid rain', or 'hole in the sky', or even 'nuclear winter' are enough to make people worried about what human activities are – or may be – doing to our planet. But 'greenhouse effect'? Surely, greenhouses are beneficial products of human ingenuity, that enable us to grow plants at times and in places that would otherwise be impossible. There has been, as some climatologists have expressed it, a 'lack of a dread factor' where the greenhouse effect is concerned, and no 'smoking gun' to implicate global warming in disasters that have befallen humankind. That may have changed in the late 1980s, as a series of extreme climatic events, including drought in the US Great Plains, made the public aware that something odd is happening to our climate. The run of hot years in that decade, whether or not they were related to the repeated El Niño events, seems to have sparked social awareness and the first stirrings of international action on the greenhouse effect, in the same way that the discovery a few years earlier of the 'hole in the sky' over Antarctica brought home the reality of the threat to the ozone layer.

It's worth recalling just a few of those disasters that seem to have had a wonderful effect in focusing attention on the problem. In the European summer of 1987, for example, the central and eastern Mediterranean experienced a dramatic heatwave, with temperatures soaring above 43°C (110°F), too much for many of the sunseeking tourists from further north to bear. The pattern was repeated in 1988, and throughout the heartlands of Eastern Europe, Siberia and North America new records for highest temperatures were set several times in the space of a few years. But while the Mediterranean baked in 1987, the same pattern of winds that kept clouds away from the region brought unseasonable rain to Britain. It was the same in 1988 – killing heat in Greece and southern Spain; the hottest July day in Yugoslavia since records began; Moscow baking in 27°C (82°F) warmth; and London a modest 19°C (66°F), colder than Copenhagen, Oslo, Helsinki or any other capital city in Europe except Amsterdam. Further east, floods in Korea and Bangladesh made headlines in 1987, and Bangladesh was hit even harder in 1988. In North America, the drought of 1988 was merely the culmination of four years of below average rainfall.

But then again, the 1980s have also seen a run of severe winters in many parts of the northern hemisphere. The pattern is not one of smooth and uniform change in temperature, or rainfall, but of seemingly erratic extremes of different kinds occurring in different parts of the world at different times. Some winters have been very cold; but some summers have been very, *very* hot, so that a modest upward trend in annual average temperatures masks some violent swings to either extreme. Researchers such as Mick Kelly, at the University of East Anglia, warn that this disruptive pattern of extreme weather variations is exactly what should be expected as the anthropogenic greenhouse effect begins to alter world weather patterns. We are, they say, in for a rough ride before the disturbances to the climate balance caused by human activities work their way through the system and a new stable pattern becomes established. Indeed, as long as we keep adding to the burden of greenhouse gases in the atmosphere, a new balance *cannot* be established, and the disturbed pattern of extremes will continue, superimposed on the rising trend of temperatures. This is well worth keeping in mind, since most projections can only tell us what the world might be like *after* equilibrium is established at a level of 3°C, say, above present temperatures. The unpleasant truth is that if these temperatures are reached in, for example, the 2020s, they will not mark the end of the rise, or even a temporary stopping place, but simply a passing landmark as the world continues to warm. In 2030, we may be living in a world that is 3°C warmer than today, and still subject to erratic fluctuations of extreme weather.

By definition, the occurrence of specific erratic extreme events cannot be predicted; but the chance that killing heat will hit Greece or disastrous floods will strike Bangladesh might be assessed in a statistical sense, and used as a basis for disaster planning. Indeed, it isn't only planning for disasters that benefits from this kind of study

– more routine requirements, such as the maximum likely demand for electric power, also depend on the extremes of temperature encountered in a country or a region each year. Climatologists are beginning to include estimates of such impacts on society in their scenarios of the greenhouse warming. But by and large their computer simulations are still only able to provide a broad picture of the kinds of climatic changes we will see over the next ten or twenty years.

The smoking gun

Jim Hansen and his colleagues in the GISS group have made the firmest statement of any climate researchers that the greenhouse warming is now upon us. Putting their reputations where their mouths are, they have projected the likely rise in temperatures over the next few decades, and have also estimated how the likelihood of extremely hot years will increase in different parts of the United States in the 1990s.

The GISS team start their computer runs with conditions appropriate for 1958, when the Mauna Loa monitoring of carbon dioxide began. They use three different scenarios for the growth of the anthropogenic greenhouse effect from that date into the twenty-first century. The first (Scenario A) assumes that emissions of carbon dioxide and other trace gases discussed in Chapter Six continue to grow unchecked – 'business as usual'. This involves exponential growth, like compound interest, with the same *percentage* rise each year (1.5 per cent for carbon dioxide*) involving a bigger increase in input of greenhouse gases to the atmosphere in real terms. The second scenario assumes that release of these gases continues to increase, but by the same amount each year, not the same percentage (Scenario B). As population continues to grow, this implies a reduction in the amount of each gas released per head of global population each year. In particular, it implies less use of fossil fuel per head. As concern about the greenhouse effect increases, the most realistic scenario on a timescale of ten to twenty years probably lies somewhere between these two projections. The third scenario is essentially the same as Scenario B up to 2000, but assumes that from then on emissions are held

*Remember that growth of carbon dioxide emissions over the past century has actually been close to four per cent.

steady at the 2000 level (Scenario C). Allowing for the effects of other greenhouse gases, this would involve cutting fossil fuel use to half its present level, and does not seem very plausible. The three emission scenarios are represented in *Figure 9.1*, where they are compared with the 'benchmark' of a doubling of the carbon dioxide equivalent in the atmosphere above its pre-industrial level, and with the carbon dioxide equivalent projected by Ramanathan's team for 2030.

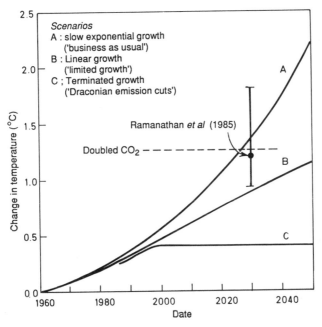

Figure 9.1
The three scenarios used by the GISS team in their forecasts, with an indication of the temperature increase that each would produce, without *allowing for feedbacks. The level corresponding to a doubling of the pre-industrial concentration of carbon dioxide is shown for comparison, together with the estimate by Ramanathan and colleagues of the carbon dioxide equivalent (including other traces gases) likely to be reached in 2030. To allow for feedbacks, these temperatures should be roughly doubled, as I have done for figure 6.6.*
(Source: James Hansen/GISS)

To turn these smooth curves into a more realistic representation of how the climate might vary over the next twenty years, the team used a GCM which takes account of many feedbacks including snow and ice cover, interactions between the atmosphere and the oceans, and cloudiness. It can also allow for the presence of stratospheric particles produced by volcanic eruptions,

201

temporarily screening out some sunlight and cooling the surface of the planet; and even for thermal inertia (but estimates of this lag beyond 2000 can only be approximate). To test how this computer model behaves, they set it up with the composition of atmospheric greenhouse gases appropriate for 1958, and let it run for a hundred simulated 'years'. The global average surface temperature in this control run varies from year to year and decade to decade almost uncannily like the actual temperature variations of the real world, but these are simply random fluctuations produced by the interactions between different parts of the model. About halfway through the control run, average 'temperatures' drop by about 0.4°C, simply because of random changes in the way the model allows features such as ice cover and cloudiness to vary (*Figure 9.2*).

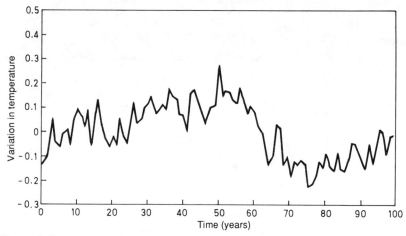

Figure 9.2
The GISS 'control run', showing the range of natural variability in their GCM.
(Source: James Hansen/GISS)

The broad features of the control run look realistic enough to suggest that it is a good guide to the greenhouse effect operating in the real world. The size of the random fluctuations, both in the model and in the real world, means that any variation up to 0.4°C on a timescale of a couple of decades should be treated with caution, since it *might* be just a random change.

When they used this GCM to simulate how climate might change in the near future under each of their three scenarios, the GISS researchers made Scenario A as extreme as possible by leaving out any hypothetical influence of future volcanic activity. They just combined the steadily rising top curve from *Figure 9.1* with the

GCM – a model which projects an equilibrium global warming of 4.2°C for a doubling of the effective carbon dioxide concentration in the atmosphere. For Scenarios B and C, however, they were a little bit more subtle, allowing for the hypothetical temporary cooling influence of large volcanic eruptions in 1995 and 2015, matching eruptions of Mount Agung from 1963 and El Chichón in 1982. The results of all this work are shown in *Figure 9.3*. The model does a fairly good job of simulating the actual run of world mean temperatures from 1958 to 1988, and then heads off into regions unknown. For comparison, Hansen and his colleagues have marked a range of temperatures which correspond to the warmest conditions of the present interglacial, the Holocene warm period of six thousand years ago, and the warmest times of the previous interglacial, the Eemian, 125,000 years ago. Even the unrealistic Scenario C envisages the world warmer than it has been for six thousand years by 2000, and staying that warm; both the other scenarios envisage the world warmer than it has been for 125,000 years (perhaps a lot longer) by the early twenty-first century, and getting warmer.

In their scientific paper presenting these calculations, published in the *Journal of Geophysical Research* in the summer of 1988, the GISS team suggested a test for the anthropogenic greenhouse effect. If global mean temperature 'rises and remains for a few years above an appropriate significance level, which we have argued is about 0.4°C,' they said, then this 'will constitute convincing evidence of a cause and effect relationship, i.e., a "smoking gun", in current vernacular'. But those words had been written months before the journal appeared, when temperature figures for 1987 had not yet been fully compiled. The projection was overtaken by events even before it appeared in print. In his testimony to the Senate on 23 June 1988, with full figures for the record-breaking warmth of 1987 and the first months of 1988 before him, Hansen said that this benchmark had indeed been reached. The observed global warming was already some 0.4°C above the 1951–1980 mean, a fluctuation which might occur by chance with a probability of just one in a hundred. This was the basis for Hansen's widely publicized comment that he was 'ninety-nine per cent sure that current temperatures represent a real warming trend'.

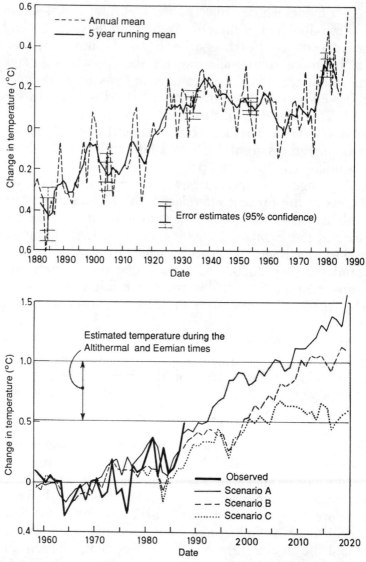

Figure 9.3
Top: *The trend in global mean temperatures since 1880.*
Bottom: *The warming implied by the three GISS scenarios (described in the text). Scenario B is the one that the GISS researchers regard as the most plausible, and includes short-lived cooling caused by the eruption of a hypothetical volcano in the 1990s. The model runs start from 1958 and allow for the build-up of greenhouse gases; they closely match the actual rise in temperature to date, and indicate how the world is soon likely to be warmer than it has been for well over a hundred thousand years. The temperature changes (relative to 1958) indicated here represent actual temperatures allowing for the thermal inertia of*

204

Into the nineties

How soon will the effects of this warming be noticeable to the person in the street? The first important point is that warming in the 1990s may be even faster than the model suggests. Because it starts up in 1958, it takes no account of any inbuilt warming that was already present because of the build-up of greenhouse gases prior to 1958.* The thermal inertia effect, mentioned earlier, makes it quite likely that there is half a degree Celsius or so of warming still to come from these past emissions (if doubling carbon dioxide causes a 4°C rise). But even without including this, average temperatures in the 1990s seem likely to rise, compared with the 1950–79 period, by about the same amount as the size of typical variations from year to year between 1950 and 1979. In the 1990s, there will still be fluctuations from year to year, but the average level will be about the same as the *warmest* years of those past decades. By the 2010s, the entire world will have warmed by much more than this, so that even the cold years will be warmer than the warm years of the 1950s and 1960s. More frequent occurrences of extreme hot summers are likely, said the GISS study written before the summer of 1988 hit the US, to be the most noticeable feature of the changes in the 1990s.

In one example of how such changes can be projected, the GISS team looked at the frequency of hot summers in different parts of the US. They defined the ten warmest summers in the period from 1950 to 1979 as 'hot', the ten coolest as 'cold' and the middle ten as 'normal'. On this definition, over that time interval the chance of a hot summer was one in three. For Washington, DC, in the 1990s, the chance of such a hot summer exceeds fifty per cent in all three scenarios, and is more than seventy per cent in Scenario A. Hansen and his colleagues explain what this means in

*And because the projections start from 1958, all the temperature increases in the scenarios are relative to the 1950–79 average. The world had by then already warmed slightly because of the accumulation of greenhouse gases since the early nineteenth century, but instead of talking about the rise since the time when the carbon dioxide concentration was a mere 280 ppm, the GISS team, logically enough, prefer to talk of the rise since 1958.

the oceans; for a doubling of the carbon dioxide equivalent, the equilibrium increase in temperature for this GCM is 4.2°C. So even if all greenhouse gas emissions halted in 2020, on this scenario the world would still have a further warming of 3°C or so built in to the system.
(Source: James Hansen/GISS)

terms of the throw of a six-faced die. In the 1960s, such a die could be used to represent the changes of getting a hot, cold or normal summer by painting two faces red, two blue and two pink. In the 1990s, the same die could still be rolled to estimate the chance of a hot summer – but now *four* of the faces will have to be painted red. There would, though, still be one blue face on the die – an occasional summer as cool as the coldest of the 1960s could still happen during the next decade.

The computer models can also be used to indicate how often temperatures are likely to rise above some particular value. During the climate reference period from 1950 to 1979, Washington experienced, on average, thirty-four days a year when the temperature rose above 32°C (90°F); according to Scenario A, that will rise to forty-two days a year in the 1990s, and fifty days a year by the

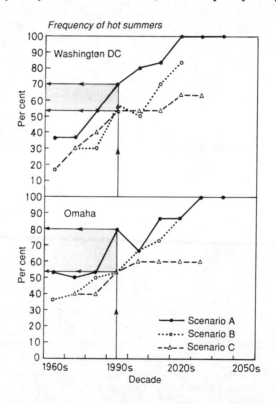

Figure 9.4
The increase in frequency of hot summers in different parts of the United States, according to the GISS scenarios. Summer heat that used to come only three times a decade will be at least twice as common in the 1990s, and may become the bottom *of the temperature range by the 2020s.*
(Source: James Hansen/GISS)

2010s. The number of such hot days in New York will increase from fourteen to eighteen a year as we move into the 1990s, and in Atlanta, Georgia, the increase is from thirty-two to forty days a year.

Farmers are especially interested in the likelihood of a run of consecutive days with high temperatures, because this can have a profound impact on the life cycles of crop plants. A run of five consecutive days with temperature above 35°C (95°F), for example, is extremely damaging to corn (maize). Between 1950 and 1979, at Omaha, Nebraska, there were three years in each decade with a run of such extreme temperatures. This increases to five years during the 1990s in Scenario A, and nine years per decade by the time the carbon dioxide equivalent of the atmosphere has doubled compared with 1880. By then, it will be impossible to grow maize in the region around Omaha.

But farmers are just as much concerned about drought as they are about high temperatures, although the two often go hand in hand. The records show that since the middle of the nineteenth century the region of the northern hemisphere between 35°N and 70°N (which covers most of Europe and America north of Los Angeles and Oklahoma) has got slightly wetter, while the region from 5°N to 35°N (including Central America, the bulge of Africa and India) has got slightly drier. The equatorial belt itself shows no significant change. In broad terms, this is exactly what should happen when the world begins to warm. Evaporation increases from regions near the equator, and the moisture in the air is carried to higher latitudes, where it falls as rain or snow (these figures are annual averages, so winter rain or snow may compensate, or overcompensate, for summer droughts in some places). When water vapour condenses, it gives up the same amount of heat that was required to evaporate it in the first place, so as well as making the higher latitudes wetter this process helps to make them warmer, transporting heat from the tropics poleward – one reason, along with the ice feedbacks already mentioned, that the higher latitudes warm more than the tropics when the anthropogenic greenhouse effect gets to work. In fact, although this data was only gathered together and published in 1987, the broad trends had been predicted in the early 1980s by GCMs developed by Syukuro Manabe and his colleagues at the Geophysical Fluid Dynamics Laboratory (GFDL) at Princeton University. The same models give us another view of the world of the 1990s. Where the GISS models concentrate on temperature changes that we can expect in the 1990s and beyond, the GFDL simulations also tell us how the rainfall belts are likely to shift.

These models are based on the global weather patterns for a world in which the carbon dioxide equivalent of the atmosphere is twice its pre-industrial level. One interesting feature of these studies is that doubling the carbon dioxide greenhouse effect produces almost exactly the same change in temperature and weather patterns as increasing the amount of heat arriving from the Sun by two per cent. Solar variations on a timescale of decades and centuries actually amount to no more than a fluctuation of about 0.1 per cent, over the Sun's eleven-year cycle of activity. But this comparison does give a very good feel for the strength of the greenhouse effect now being imposed on the Earth as a result of human activities. We are 'forcing' the weather machine in a new direction, with a push twenty times stronger than anything the Sun can do on the same timescale.

Another feature of the earliest runs with these models is that they show a drying of continental interiors in a region centred on 42°N latitude. This does not really conflict with the more recent analysis of actual rainfall trends, since it is dealing with a much more extreme disturbance of natural climate systems than we have yet achieved: by the time the carbon dioxide equivalent is doubled, the region of enhanced rainfall is certain to have moved further away from the equator than it is today. And, as we shall see, winter rainfall may be masking the increasing dryness of summers in places such as southern Spain and Greece. This latitude band encompasses most of Spain and Portugal, Italy and Greece, as well as large chunks of Nevada, Utah, Colorado, Wyoming, Nebraska, Kansas, Iowa, Missouri and Illinois. The scenario provides especially intriguing food for thought in the wake of the heat and drought of 1987 and 1988.

These early studies have been refined by the GFDL team in later years. They include studies of regional temperature variations, which broadly match those of the GISS scenarios, and show that Antarctica is likely to warm less than the Arctic region, because of differences in the ice feedback mechanism in the two hemispheres. In the north, thin layers of snow and ice lying on land can easily melt; in the south, at equivalent latitudes there is only open ocean. Thick sea ice and great, thick ice sheets over land around the South Pole cannot easily be melted at all. Because land warms more rapidly than the oceans in any case, the northern hemisphere as a whole will also warm more quickly than the southern hemisphere at first, although if equilibrium is ever established (if the build-up of greenhouse gases is brought to a halt) then the oceans, and the south, will catch up. The later simulations also identify two separate regions of summer dryness in a warmer world. The first is the region around 42°N identified in the early studies. The Great Plains of North America, and all of southern Europe both dry out, because rainfall shifts northward and because the greater summer heat causes more evaporation from the soil. The lack of moisture to be evaporated itself feeds back on the warming process to make the region hotter still. Just as a damp cloth placed over a bottle of milk in summer will help to keep the milk cool because heat is taken up by moisture evaporating from the cloth, so moisture in the soil keeps the ground (and the air above it) cool in summer as heat is used to vaporize the water. When there is less moisture, this is less effective: heat that would otherwise go into evaporating water (which might then be transported northward) goes instead directly into warming the surface of the ground. This effect dominates in southern Europe, but another factor is also at work on the Great Plains of North America and much further north. There, the early melting of snow in spring in a warmer world means that water begins to evaporate from the soil much earlier in the season. By the time summer comes the soil is already drying out, and so most of the incoming solar heat is absorbed by the ground.

The whole rainbelt shifts poleward as the world warms, not just from year to year but also between winter and summer. A region in middle latitudes (around 40 to 50°N) that is in the northern part of the rainbelt in winter would be in the southern part in summer. So as the influence of carbon dioxide and other greenhouse gases increases, farmers in that part of the world will find that they still get adequate winter rains, but that summers become increasingly dry. Indeed, because there is more moisture in the air in a greenhouse world, winters may actually become *wetter* with more precipitation as rain and snow at the same time that summer droughts become more common. But unless present reservoir capacity is greatly increased, most of this extra precipitation will be useless, and will run away off the land in spring before it can be utilized by crops. Yet again, the predicted pattern is one of extreme variations, the kind of pattern that is masked by simple measurements of annual average rainfall. The central regions of both the US and Europe can look forward to more winter blizzards in the 1990s as well as more summer droughts.

But these projections are for the kind of conditions that will exist in an *equilibrium* climate for a doubling of the carbon dioxide equivalent. In the late 1980s, experts increasingly voiced their concern not at the magnitude of these changes, but at the speed with which they will occur. If the transition were slow and gentle, then both human activities and the natural world would have time to adapt – as they have, indeed, adapted to the warming of a modest 0.5°C that occurred between 1880 and 1980. But now we are talking about a rise four times bigger in less than one third of a century – a rate more than ten times more rapid than the increase over the past century, and probably faster than any climatic change of a comparable size that has occurred since the Earth formed. Can Gaia possibly be expected to cope with that?

How fast?

The two more plausible GISS scenarios suggest that over the next thirty years or so temperature and vegetation belts will be moving poleward in mid-latitudes at a rate of between fifty and seventy kilometres per decade. That is about ten times faster than plants have ever been known to respond to climate shifts in the past. But if you are tempted to think that these scenarios, based on a GCM which produces a 4.2°C global mean warming for a doubling of carbon dioxide, are at the implausibly high end of current estimates, think again. In October 1988, the former Director-General of the UK Meteorological Office, as cautious an organization as you are likely to encounter when it comes to making climatic predictions, stunned his colleagues around the world by announcing that the latest version of the Met Office's own GCM projects that a doubling of the carbon dioxide equivalent will raise global mean temperatures by 5.2°C, with the Arctic warming by as much as twelve degrees. The doubling could happen well before 2050; the actual warming will take decades longer, because of thermal inertia. The important feature of the new findings, said Sir John Mason, is that the temperature increases predicted for high latitudes become greater as the complexity and sophistication of the models are increased. The more account is taken of changes in cloud cover and associated feedbacks – the more the models match the real world – the worse the prospect becomes.

The official Met Office response to these comments from their former boss was frosty. Different versions of the Met Office GCM give somewhat smaller warmings, they pointed out, and the figures he quoted had not yet been published and did not represent their official view, as of October 1988. Even if, however, these figures represent the wild extreme of current estimates, that leaves the GISS prediction of a 4.2°C warming for a doubling of the carbon dioxide equivalent very much back in the mainstream of respectable forecasts. And that puts their scenarios for the speed of change over the next few decades on a very secure footing. In fact, most ecologists who have looked at the problem are struggling to catch up with the latest projections of the speed of climatic change, and concentrate on the impact of a rise in temperatures of three to eight degrees Celsius over the next hundred years. The bottom end of this range, and perhaps even the top, is *less* than now seems likely, but still sufficient to wreak ecological havoc. Forests are a good benchmark of the health of Gaia. Add a couple of degrees Celsius to present temperatures, and the remaining temperate forests in the northern hemisphere start to die at the southern limit of their range. New habitats for them might be appearing further north as the climate zones shift poleward – but how can the trees get to their new 'correct' latitudes? Unlike Dunsinane Wood, they cannot just walk there. And their problems are simply symptomatic of the difficulties confronting many kinds of life.

Using what now seems a relatively modest forecast of a global warming of 3°C over the next fifty years, the world will be warming fifty times faster than it did at the end of the latest ice age. Even though you might expect a warmer world to be more amenable to life, the pace of *that* change killed many species that could not move fast enough to find new homes. Among the mammals alone, thirty-two genera became extinct. Ecologists do not fully understand what controls the distribution of species. It is clear that many factors, including climate, competition from other species and availability of food play a part, affecting each other in a complex web of interactions. If any part of the web changes, the more complex an organism is the harder it will find the task of adapting to the change. Microorganisms generally evolve quickly to adapt to a new environment, tracking the changes almost as fast as they occur. Other species, such as mammals, are slower to adapt, and while under the strain of adapting and/or migrating may come under attack from disease organisms and parasites.

Parasites generally thrive in a warmer climate. The barber worm that infests sheep in Australia, for example, reaches epidemic proportions in parts of the country in summer, but declines in winter. In southern parts of Australia it is not, at present, a threat. But as the climate warms it will become a bigger and more permanent problem. Similarly, parasites that we now think of as typically tropical will spread north and south into the temperate zones. Hookworm, a particularly nasty little beast that hooks onto its human victims (hence the name) then gets inside its unwilling host by boring through the skin, could become a major problem in developed countries of the northern hemisphere, along with the malaria mosquito.

Leaving those problems aside, however, by and large mature plants and animals are much hardier than the young. That is fine for them – they may survive as the world warms. But it means there will be no new generation to replace them when they die. Migratory species may be the first animals to succumb, because although they are able to travel long distances, they time their journeys to match the availability of food supplies along the route. Take out even one link in that chain of replenishing stops, and the migratory species goes with it.

The connections are everywhere that biologists choose to look. Increasing the speed with which snow melts in the spring, for example, has a damaging effect on wildlife. The melt-water is acidic, and under the normal conditions of recent centuries as the winter snow melts slowly it has time to sink into the ground where it is neutralized. But more rapid snowmelt will produce a flood of acidic water, too much to be absorbed by the soil, that will flow off into the rivers and lakes in a killing acid pulse at a time when the young of many freshwater species are at their most vulnerable stage, just hatching or still in the egg. The grim catalogue seems endless. But always the talk among the experts returns to the fate of the forests.

Warming at a rate of 0.4°C per decade, in line with the GISS scenarios, will shift climate zones so rapidly that forests will have to move at a rate of six hundred kilometres a century. When the world warmed at the end of the latest ice age, the fastest rate of movement was two hundred kilometres a century – for spruce. This is something of an arborial speed record; a more normal rate is about twenty kilometres a century. Margaret Davis, of the University of Minnesota, has studied four important North American trees: yellow birch, maple, hemlock spruce, and beech. By the middle of the next century, all four will die along their southern range. Under the new climatic regime, they will 'belong' between five hundred and a thousand kilometres north of their present locations, shifting beech, for example, from the southeastern US to Ontario and Quebec (and the world will *still* be warming!) Although some varieties of beech do grow today in Maine, there is little chance that those particular types will survive there when the climate changes to that of present day Georgia, as it will within the lifetime of a single tree. 'Dieback' will begin by the end of the 1990s, and will show in the form of forests full of ageing trees, with no new saplings growing up to take their place. Conifers are already being pushed out of the temperate region of North America and Europe, being replaced by broad leaved varieties. But their success seems likely to be short lived.

Perhaps we can live in a world without forests, with trees confined to artificial parks. But what will happen to agriculture while all this is going on?

Agricultural impacts

It is impossible to predict how our food supplies will be affected by the increasing strength of the anthropogenic greenhouse effect. To start with, plant biologists have only been able to study the response, in any one experiment, of one or two individual plants, not a whole ecosystem, to such environmental changes as a rise in carbon dioxide concentrations. In addition, there is no way to foresee how farming practices will change as the greenhouse effect begins to bite. So this section does not even offer scenarios for what the future might hold; instead, I want to sketch out some of the difficulties that the planners have to face in trying to come to grips with this kind of problem.

Plants need carbon dioxide, and in some ways adding more carbon dioxide to the air will be beneficial to plants – the so-called 'carbon dioxide fertilization' effect. But what is good for an individual plant may not be just what a farmer, with a field full of plants, requires. In experiments at Clemenson University, South Carolina, samples of the crop sorghum were grown in an environment with a concentration of carbon dioxide twice as high as in the atmosphere today. Individual plants grew bigger, all right; but their leaves were so big that they shaded each other. Even for individual plants, the enlarged leaves were not able to make maximum use of their potential for photosynthesis because of the shading; in a field full of such plants, although each plant might be bigger they would have to be spaced further apart to avoid even more shading. Would a field full of a smaller number of larger plants produce any more food than a field full of a larger number of smaller plants, each making maximum use of its leaf area for photosynthesis? Nobody knows. But the best way to find out might be to look at changes that are already taking place in forest ecosystems.

One effect of a higher concentration of carbon dioxide is that plants use water more efficiently. Ian Woodward, of the University of Cambridge, has studied the way in which plants lose water, through pores in their leaves called stomata, as they take in carbon dioxide. The same holes let carbon dioxide in and water vapour out; if there is more carbon dioxide in the air, you might guess that the holes could be smaller, since it is easier to absorb carbon dioxide, and you would be right. Woodward found that there is a significant difference between leaves collected from living trees today, and specimen leaves in herbaria that were collected from similar trees two hundred years ago, before the Industrial Revolution. Trees now have leaves with fewer stomata compared with their close relatives at a time when the carbon dioxide concentration of the air was seventy ppm less than it is today. If there are fewer holes in the leaves, then water should not escape from the leaves so easily. Assuming this is a response to the build-up of carbon dioxide, it means that trees, and perhaps other plants, are already using water more efficiently. Such a trend might seem to be good news for farmers, suggesting that it will reduce the need for irrigation. But in natural ecosystems it could also mean that with less water being taken up by trees there is more running over the surface of the ground and through the soil, causing erosion, leaching minerals from the soil and increasing the risk of flooding from swollen rivers.

This isn't the only evidence that trees are already changing as the carbon dioxide level increases. Bristlecone pines, that live close to the treeline on mountains in the western United States, have been putting on extra inches over the past century. These are long-lived, slow-growing species. Their growth can be monitored by drilling pencil-thin cores from their trunks, and comparing the width of the rings laid down in succeeding years. Trees at the high altitude limit of their range, struggling for survival for centuries, seem to have been succeeding in that struggle more effectively as the amount of carbon dioxide in the air builds up. And a similar effect may be at work in Britain.

The evidence emerged in a letter published in the *Guardian* on 23 December 1987, part of a heated correspondence on the subject of acid rain. It came, ironically, from a reader concerned to dismiss the notion that human activities were altering the growth of trees. Allan Mitchell, who has made a special study of the growth of trees across the British Isles, was responding to claims that trees were suffering a dramatic loss of vigour because of acid rain. Far from it, he said. The results of thirty-five years' work measuring and remeasuring seventy thousand trees of 1500 species showed that redwoods have lately been adding three to four inches to their girth each year, while in 1987 many species showed unusually large leading shoots, three to four feet in size, at the tips of their branches. Whatever was influencing these trees, said Mitchell, 'it certainly cannot be acid rain'. Could it be the build-up of carbon dioxide in the atmosphere?

Woodward says that as carbon dioxide concentrations continue to increase there should be an increase in the hardiness of vegetation, and that shrubby plants should begin to spread into areas that are now sparsely vegetated in east and southern Africa, Saudi Arabia and Australia. But this raises the possibility of another problem for agriculture. 'Weedy' plants might do every bit as well as crop plants as the carbon dioxide concentration builds up; might they, indeed, achieve more success, on farmland, at the expense of the crops?

Prognostications of this kind are complicated by the fact that different kinds of plants actually carry out photosynthesis in slightly different ways. Two particular types are important in this discussion. One family is called C3 plants, and the other C4. In each case, the label indicates the number of carbon atoms in each molecule of the main type of sugar that each plant manufactures out of carbon dioxide and water as a result of photosynthesis (the sugars are then converted, in a series of biochemical processes, into other types of molecule such as carbohydrates and proteins). This seemingly small difference actually reflects a significant difference in lifestyles of the two families. C4 plants are more productive than C3 plants under optimum conditions, but they need strong sunlight and high temperature to do well. C3 plants are less productive than C4 plants in the tropics, but come into their own at latitudes above about 45°, where the C4 species are less well adapted. In the mid-range where the two types overlap, their lifestyles are often complementary, with C3s growing during the cool, moist months and C4s growing best in the hot, dry season.

Only a few of the major food plants are C4s – maize, sorghum, sugar cane and millet. Most of the staples found on farms, including wheat, rice, potato, barley and soybean (and, indeed, virtually everything else) are members of the C3 family. When there is more carbon dioxide available, both types of plant increase their

215

photosynthetic activity, but the effect is more pronounced in C3s; at the same time, both types use water more efficiently, but this effect is more pronounced in C4s. On balance, C3s benefit more than C4s from an increase in the amount of carbon dioxide available, growing bigger and more quickly, and extending their range. In some parts of the world, this will benefit agriculture. But there are also C3 weeds, as well as C3 crop plants. In places where C4 crops such as maize (corn) are benefiting only a little from the build-up of carbon dioxide, C3 weeds may thrive and crowd out the crops.

And there is another problem. David Lincoln, of the University of South Carolina, is one of several researchers who have studied how the composition of leaves, as well as their size, changes when plants are 'fertilized' with extra carbon dioxide. The whole biochemical factory inside the plant adjusts to the increased availability of carbon dioxide in such a way that a greater proportion of the carbon taken up during photosynthesis is stored in the form of carbohydrate, and a smaller proportion is converted into protein. So, in terms of protein content, the big leaves are less nutritious than the smaller leaves of similar plants grown in an atmosphere with less carbon dioxide. This may have repercussions in terms of human food supply; but the main thing that the experts are worried about now is that it makes the leaves less nutritious for insect pests. Why should they care if the insects don't have sufficient protein? Because the response of an insect herbivore to a decline in the amount of protein it gets from each bite of leaf is to eat more leaf! In an unbalanced, rapidly changing situation this could result in severe damage to crops in the fields in some years.

All of this, though, takes no account of the climatic changes that will be going on at the same time as the increase in carbon dioxide concentration. Increasing the availability of carbon dioxide may be beneficial to plants, *other things being equal*: but other things will not be equal in the hothouse world of the twenty-first century. Researchers from the University of Nottingham and the University of Bristol recently tested growing winter wheat under conditions in which either the amount of carbon dioxide available was doubled, or the temperature was increased by 4°C, or both. With a doubling of the carbon dioxide concentration alone, yields increased by about twenty-five per cent; but with a concurrent increase in temperature, the crop passed through its various stages of development more quickly, the grain matured earlier and each grain was smaller, so the total yield was less impressive. For a temperature increase of 4.5°C and a carbon dioxide doubling, the two effects cancelled out. The SCOPE 29 study suggests that in present day core areas of the mid-latitude grain-growing regions of North America and Europe, the effect of an increase in temperature of 2°C will be to reduce crop yields by between three and seventeen per cent. This is the effect of the temperature increase alone; decreases in precipitation will make the situation worse in some places, while in regions on the edge of continents, such as the British Isles, increased rainfall might ease the problem. Some crops in some parts of the world will do better as the greenhouse effect becomes more potent; other regions will suffer agricultural decline. One of the key tricks in making any plans for life in the global hothouse is to work out which regions will benefit, and which will suffer. The computer models are not yet good enough to make forecasts of this kind unaided. But there is another way to get a picture of the changes in regional climate corresponding to a warmer Earth: study the pattern of regional rainfall and temperature changes by comparing cold years from recent history with years when the world actually *was* warmer.

Several groups of researchers have used variations on this idea to get an insight into climatic changes of the twentieth century. One study, by a group at the University of East Anglia, looked at temperature records from many northern hemisphere sites for the fifty years from 1925 to 1974 and picked out the five warmest years (1937, 1938, 1943, 1944 and 1953) from that interval, for comparison with the five coldest years (1964, 1965, 1966, 1968 and 1972). They then combined the regional rainfall and temperature data from each of the five warm years to give a 'warm composite', and from each of the five cool years to give a 'cool composite'. By comparing the two composites, they found that when the world is warmer maximum warming occurs in the continental interiors at high latitudes. This is very much in line with computer modelling scenarios, but provides even more detailed regional information.

For these particular 'real world' scenarios, the overall temperature difference between the two composites was 1.8°C in winter and 0.7°C in summer; but over the whole 'year' a region from Finland across Russia and Siberia to 90°E warmed by up to 3°C, much of North America showed increases of one to two degrees, and some regions actually showed a fall in temperature. The warm composite also showed an overall increase in precipitation of one to two per cent – but there was a *decrease* over much of the US, Europe,

217

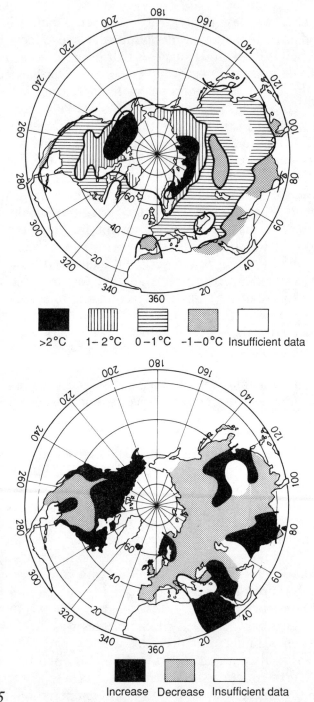

>2°C   1-2°C   0-1°C   -1-0°C   Insufficient data

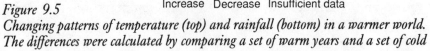

Increase   Decrease   Insufficient data

Figure 9.5
Changing patterns of temperature (top) and rainfall (bottom) in a warmer world.
The differences were calculated by comparing a set of warm years and a set of cold

218

Russia and Japan, balanced by bigger than average increases over India and the Middle East. The broad picture can be seen in *Figure 9.5.*

Of course, this pattern of changes may not be exactly the pattern that will develop as the greenhouse effect takes a grip, and it deals with changes much smaller than those we can expect in our lifetimes. But by combining studies of this kind with computer simulations and experience of growing crops under different conditions today, it is possible to begin to get a picture of the changing agricultural scene that will develop as the world warms in the 1990s and beyond. Just such a study was carried out in the 1980s by a large team of researchers at centres around the world, for the United Nations Environment Programme and the International Institute for Applied Systems Analysis. The study took four years to complete, and its findings were published in two hefty volumes, totalling 1,640 pages, in 1988.* I shall pick out just a few of the conclusions as further food for thought.

The study is based on the now standard scenario of a doubling of the carbon dioxide equivalent in the atmosphere, compared with pre-industrial times, over the fifty years from 1980 to 2030, with a resulting rise in global mean (equilibrium) temperatures of 3.5°C. The researchers found that even 'beneficial' shifts in climate could pose problems, with huge surpluses of some crops being produced in some parts of the world. Rice today, for example, is grown on four million farms in Japan, and the farmers are subsidized by government intervention which has kept prices in the late 1980s three to four times higher than those on the world market. As it becomes easier to grow more rice on those same farms, and as world prices fall because other regions can also grow more rice, the subsidy and the resulting 'rice mountain' could become important economic and political issues.

Just the opposite kind of problem will hit the prairie wheat belt of the US and Canada, with a return to dry and windy dust-bowl conditions in the 1990s. This scenario from the report, written before the drought of 1988 but published after it hit, looks strikingly prescient. As a case study, the researchers looked at the

*Edited by Martin Parry, Timothy Carter and Nicolaas Konijn; see Bibliography.

years from the twentieth century. High latitudes (nearer the pole) warm most; continental heartlands dry out, but some regions nearer coasts become wetter. (Source: CRU, University of East Anglia)

219

impact on one Canadian province, Saskatchewan. They suggest a dramatic setback to farm production, with the loss of 1,200 jobs in the agricultural sector and 2,600 jobs in other sectors, related to a decline of Can$ 275 million a year in farm income and almost as much in other sectors of the economy – for one province, not even the whole country.

For decades, the North American plains have produced a surplus of grain, and one of the biggest customers for that grain has been the USSR. But the same climate shift that will bring dust-bowl conditions to North America will boost agriculture in the Soviet Union. In the main farming region around Moscow, yields could go up by almost fifty per cent, provided that steps are taken to introduce varieties that will thrive under the changing conditions. Imagine a world, perhaps only twenty years away, in which the Soviet Union is a major exporter of grain, and the United States has to go cap in hand to be allowed to purchase food to keep its people from going hungry. Even if there were *more* agricultural production in the world as a whole, it would be a very different place politically.

Countries such as Finland and Iceland, in the cold part of the world, will also benefit. But semi-arid regions of Brazil, Australia, India and Africa will suffer. Within fifty years (even allowing for thermal inertia), the climate of Finland will be like that of northern Germany today; the Leningrad area will become like the western Ukraine; southern Saskatchewan will be like northern Nebraska; southern England will be like the south of France (but probably wetter). The warming trend will bring about geographical shifts in growing patterns of several hundred kilometres for every degree of temperature change, says the report. 'It will also increase the frequency and magnitude of severe shocks to agriculture from major floods, persistent droughts, soil erosion, forest fires and crop pests', in the words of Martin Parry, of the University of Birmingham, who directed the team of seventy-six scientists from seventeen countries involved in the project, the first international study of this kind.

The details are less important than the broad message, which can be summed up in one word – upheaval. It doesn't matter that some regions will benefit from the changes, in theory, if agricultural activities are, in practice, unable to change rapidly to take advantage of the new opportunities. The pace of the change and the inertia of political, economic and agricultural systems may outweigh, in the short term, the very real benefits that the greenhouse effect might bring to some parts of the world. An optimist might look at the projections for 2030 and conclude, rightly, that this would be a world better suited for humankind, where more food could be grown and a larger population might be fed adequately. But it is getting from here to there that is the problem. Indeed, it always reminds me of the old joke about a traveller in Ireland who asks a native the way to Dublin. 'Well now,' says the local, scratching his head, 'sure, and if I was going to Dublin, I wouldn't be starting out from here.' Like that traveller, we have no choice. We are being carried willy-nilly into the hothouse world of the twenty-first century, and we can only start out from where we are now, with bizarre global agricultural and economic systems that produce huge surpluses in Europe, for example, while millions starve in Africa.

The upheaval caused by the greenhouse effect might be the last straw that sends the whole house of cards tumbling down; or it *might* provide the incentive to open our eyes to reality and force nations to work together to solve the problems it poses. Even then, though, remember that there is no magic switch to turn the greenhouse effect off if we do arrive in 2030 at a comfortably warm, fertile world. The warming will continue to increase as long as we continue to add to the burden of greenhouse gases in the atmosphere. Just how far *could* it go, if not in our lifetimes then in the lifetimes of our children?

How far?

If the computer models are not yet good enough to give us more than a broad outline of the shape of climate to come, and our planet is already warmer than it has been for a thousand years, where can we look to find an analogue of the situation we will encounter in the twenty-first century? The only answer is to look back in time, far beyond a thousand years, to epochs when the world was warmer still. The warm period after the latest ice age, the optimum (or Hypsithermal) is an obvious place to start. *Figure 9.6* shows one reconstruction, based on a variety of geological evidence, of the broad pattern of rainfall differences of six thousand years ago, compared with rainfall of the mid-twentieth century. (At that time, the average temperature of the globe was about 1.5°C higher than that of the 'normal' period from 1950 to 1979.) But on the way 'back' to those conditions, the world will first pass through a state rather like that of the little optimum, which was at a peak

221

between about 900 and 1050 AD, with temperatures a degree or so above the climatic norm.

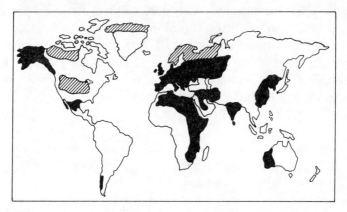

Figure 9.6
During the warmest period of the present interglacial, between about four thousand and eight thousand years ago, some parts of the world (black) were wetter than today, while others (shaded) were drier. No information is available for the blank areas.
(Source: Will Kellogg, US National Center for Atmospheric Research)

The best summary of what this, and further, warming might mean comes from Hermann Flohn, who carried out a study of the implications for the International Institute of Applied Systems Analysis (IIASA) at the end of the 1970s. The little optimum, or medieval warm period, was the first of four successively warmer (and successively more remote) periods of Earth history that he described in detail as analogues of the phases the world will pass through in the twenty-first century. Indeed, the temperature of the globe is now rising so rapidly that the first of these scenarios, for the little optimum, may be a guide to the climatic trend of the rest of the *present* century, the decade of the 1990s.

During the little optimum, forests in Canada advanced a hundred kilometres north of the present timberline, and their counterparts in Europe ranged further north and higher up mountains than they do today. There were frequent droughts everywhere in Europe south of 60°N, the latitude of Oslo and Leningrad, but the northern Sahara was wetter than it is now, and horse caravans could cross regions that are now passable only by camel train (or mechanical vehicles). There is very little evidence available about conditions in North America at that time, but what there is indicates a combination of higher temperatures and less rainfall, while in the southern

222

hemisphere New Zealand experienced severe drought and extensive forest fires. The evidence is interpreted as indicating a shift of the main storm path, the track of rain-bearing cyclonic systems, up to 60–65°N in the northern hemisphere (with a comparable shift in the south), leaving regions further from the pole dry in summer but suffering blizzards in winter under the influence of persistent 'blocking' high pressure systems. This is the best forecast available for the conditions that are likely to return in the 1990s (if they are not already upon us), bringing serious shortages of fresh water in Spain, California and the Near and Middle East. Writing in 1980, Flohn said that a person now twenty years old might see food shortages and population migrations in areas of high living standards before reaching mid-career; that hypothetical twenty-year-old would now be entering his or her thirties. But the good news is that the next phase of global warming might be more favourable to humankind than the first turn of the screw.

The climatic optimum, or Hypsithermal, following the latest ice age was the warmest epoch of at least the past seventy-five thousand years. Global mean temperatures then were about 1.5°C higher than those of the climatic reference period, and such temperatures are likely to be reached again in the first decade of the twenty-first century. But remember that the world will not then be in equilibrium, as it was more or less, six thousand years ago. The world – Gaia – might be trying to shift herself into the climatic patterns appropriate for such temperatures, but within a decade – before she has had time to adjust – temperatures will have increased still further. Nevertheless, let's look at what Gaia might be trying to move towards within twenty years.

Six thousand years ago forests in western Canada and in western Siberia extended as much as three hundred kilometres further north than they do today, and summer temperatures at these latitudes were two to three degrees higher than the average from 1950 to 1979. At the peak of the warm period, the waters between Taiwan and Japan were as much as 6°C warmer than today, and with ocean temperatures higher it is no surprise that the world was wetter, overall, than we have been used to, with more humidity in the air and more rainfall. The Sahara and deserts of the Middle East were wetter; and they seem to have shrunk at both their northern and southern margins, so this cannot simply have been because of a bodily shift of climate zones northward. Flohn says that although the tropical rain belt did shift northward, the influence of the polar ice cap still dominated in winter to produce the typical 'Mediterranean' pattern of strong winter rainfall further north; there may even have been more winter rain than today, because the seas were warmer. So the Sahara, and other deserts, were squeezed from both sides.

The same pattern applies across the arid belt from Mauretania in western Africa to Rajasthan in India. On the edge of the Tharr desert in India, where in recent decades rainfall has averaged 250 millimetres a year, they had between twice and three times as much rain a year during a long moist interval from about 10,500 years ago to 3,600 years ago. All of this might seem like good news for regions of the Earth that have been struck by terrible drought and famine in recent years, such as the Sahel and Ethiopia. Almost paradoxically, as Flohn points out, a warming of 1°C seems to be more damaging for human agricultural activities than a warming of 1.5°C. But Flohn also sounds a cautionary note. Even if the humidity of the Hypsithermal does return, he says, it may not mean that these desert margins become agriculturally productive once again. Human activities have destroyed vegetation and encouraged the spread of desert in recent decades, eroding the soil, depleting ground water and increasing the salt content in the ground water. Reversing this trend will require not only a large increase in rainfall, but a deliberate and sustained human effort to repair the damage. A modest global warming is unlikely to produce sufficient rainfall to halt, let alone reverse, the process of desertification unless we mend our ways regarding the destruction of these desert margins.

Within thirty years, unless we mend our ways regarding the emission of greenhouse gases, the world will be warmer than it has been at any time during the present interglacial. That means we have to look back to the previous interglacial to find an analogue of the conditions we are likely to experience. The further back in time we probe the geological record, the harder it is to interpret the details of past climate, and as we saw in Chapter Eight there is even some argument about just how warm the Earth was at the peak of the previous interglacial, some 125,000 years ago. But climatic data from mid-latitudes in Eurasia and North America indicate that the temperatures in those parts of the world were two to three degrees higher than in the same regions between 1950 and 1979. Judging from the computer models, and from the differences between warm and cold years in the twentieth century, the average temperature of the whole globe would have been a little less different from the present day average, while the polar regions would have been relatively warmer, compared with their present state. Sea level, though, was between five and seven metres higher than today, almost certainly, as Robin argues, because the Antarctic ice sheet was *thinner*, not because it covered a smaller area.

The high sea level had a profound effect on both the geography and the climate of northern and eastern Europe. Scandinavia was an island, with an ocean channel linking the Baltic Sea to the White Sea in the north, cutting through the low-lying lake district of the border country between Finland and the USSR. Such an increase in sea level is not expected in our lifetimes, even when the world warms by 3°C or more, since the Antarctic ice sheet is both thick and stable today. This makes it almost impossible to find an analogue for such a modern version of the warmer Earth, because the presence of all that water 125,000 years ago produced different patterns of rainfall and temperature. The sea penetrated, for example, far into Siberia along the flood plains of the Ob and Yenisey rivers, bringing moisture and rainfall to the northern heartland of what is now the USSR.

Trees are always a good indicator of climatic patterns, and their remains do indeed show that oak, linden, elm and hazel extended much further north then than they do now. But perhaps the best perspective on what such a climate shift means comes from southern England. There, hippopotamuses, forest elephants and lions were native inhabitants of the region around the present day river Thames 125,000 years ago; their remains have been found in gravels, laid down during that interglacial, beneath Trafalgar Square. Such animals could, in fact, survive at these latitudes today (and some of them do survive without much protection from the elements in safari parks). Part of the reason why there are no lions in England is that the spread of human civilization in Europe during the present interglacial has rather restricted the opportunities for lions to migrate northward. But this is only part of the story; there is a difference between being able to survive at the latitude of London and flourishing there. Flohn says that these animals were indeed flourishing there 125,000 years ago, and that their presence was 'a remarkable feature of this warm climate'.

Hippos in Trafalgar Square. The temperature increase needed to provide a suitable climate for that to happen again will occur by the 2020s, unless emissions of greenhouse gases are drastically reduced. Controls on such emissions might slow the warming, but are unlikely to halt it even at these levels. The implications of any further warming are impossible to predict in any detail, but

225

they might be upon us well before 2050, and are almost certain to be upon us before the end of the twenty-first century. If the details are hard to predict, though, one big change stands out as inevitable. The last of Flohn's four scenarios really does put the impact of human activities on Gaia in perspective – a very disturbing perspective even if you do not expect to be around to witness it, and hope (?) to leave the problem to your children. A global warming of 4°C, which could happen by the 2030s and is almost certain to happen within a hundred years, will, says Flohn, be sufficient to remove *all* the ice over the North Pole and the Arctic Ocean.

A lopsided climate

The idea of a lopsided planet with an icecap over one pole but not over the other seems bizarre to us simply because there have been icecaps over both poles throughout recorded human history – and, indeed, for millions of years. But the geological record shows that during most of its own history the Earth has had no ice at either pole. Polar caps only form when the continents drift into such a position that they block the flow of warm water to high latitudes. Even then, there is usually only one polar cap during an ice age; the conditions that have persisted during the emergence of the human line have been unusual, and probably unique, during Earth's history.* At present, the Antarctic icecap is very stable, and for all practical purposes can be regarded as a permanent feature of our planet. But the Arctic icecap, Flohn points out, is not only a much less secure feature of the planetary environment, but wasn't there at all for millions of years after the Antarctic icecap formed.

The present broad pattern of geography and global climate began to emerge as the continents moved slowly (at a rate of a couple of centimetres a year) into their present positions. Continental drift continues, and in millions of years' time the geography of the globe will be different from the pattern we know, but that need not worry us now. The first major climatic landmark on the road to what we think of as the 'normal' pattern of the world occurred between twelve and fifteen million years ago, when a large Antarctic icecap formed. At that time, the temperature

*Which may not be a coincidence; see *Children of the Ice*, by John and Mary Gribbin (Blackwell, 1990).

226

of the ice was probably only a little below freezing point, but this was enough to see the polar cap established as a 'permanent' feature which has persisted, with some fluctuations in size, right down to the present day. The first mountain glaciers in Alaska appeared a little later, and after a series of long, slow climatic fluctuations the Antarctic experienced a period of intense cold between about five and 6.5 million years ago. Then, the volume of Antarctic ice was even bigger than it is today, and sea levels were lower worldwide – so low, in fact, that the shallow strait between Spain and Africa dried up, and the Mediterranean Sea itself evaporated. But even then, five million years ago, although the south experienced an ice age worse than anything since, the Arctic Ocean was still free from ice. The 'permanent' northern icecap that seems so natural to us may only have appeared as recently as 2.5 million years ago, and certainly had not appeared by three million years ago. For at least ten million years (from around thirteen million years to three million years ago) the Earth existed with one glaciated pole and one free from ice.

If it happened once, it can happen again. Calculations and computer models show that a warming of 4°C is enough to remove the ice from the Arctic Ocean (but not immediately from the Greenland icecap, which is a thicker sheet grounded on land), and the geological record shows that the Earth can exist in such a stable but lopsided state, with ice over the South Pole but not over the North. What can the geological record tell us about the climate of the Earth under such lopsided conditions?

The analogy cannot be exact, because the continents have shifted slightly even during the past three million years – the Atlantic Ocean, for example, has got about fifty kilometres wider. Even more importantly, major mountain ranges have grown dramatically during that time. The Tibetan Plateau has been elevated from sea level to its present altitude, four to five kilometres above sea level, and this is now a major influence on the easterly winds of the monsoon system, which must have been quite different three million years ago. Even so, the broad features of the climate from the time when the Earth had only one ice-covered pole are striking enough.

You can begin to get a feel for the size of these changes by looking at their effect on the Arctic itself. It doesn't take a very large global warming to melt the ice, because it is only a thin layer – a few metres thick – floating on the water (so thin, in fact, that submarines have been able to surface close to

227

the pole itself, crashing through the ice as they do so). When this floating ice melts, it has no effect on sea level. But it does reveal a darker surface, energy-absorbing water instead of heat-reflective snow and ice, and this makes it very difficult to re-form the Arctic icecap once it disappears. A warming of just 4°C will do the job of getting rid of the icecap, but temperatures would then have to fall at least as much as this *below* those of today (eight degrees down from the temperature at which the icecap melts) in order to bring it back again.

The average surface temperature of the air over the Arctic in winter today is about –34°C. If the world warms by just 4°C, so that the Arctic ice melts, the average temperature of wintertime surface air over the Arctic will increase to +4°C – a rise of 38°C compared with the present day. By any standards, this is a major alteration in one of the key components of the weather machine, and is on a completely different scale of importance from any other aspect of the anthropogenic greenhouse effect. At present, the surface of the Arctic icecap is so cold in winter that the air above it, which has come in from lower latitudes, is actually warmer than the surface. Just above the surface, temperatures increase with altitude up to about –20°C before beginning to cool again on the way up to the stratosphere. This inversion, as it is called, inhibits convection over the Arctic and has an important influence on the circulation of the whole northern hemisphere. Under ice-free conditions, the temperature of the atmosphere falls all the way from the surface to the stratosphere, completely changing the pattern of convection. Averaging over the whole troposphere above the Arctic, Flohn calculates a warming of 11.5°C in winter and about 3°C in summer. Using averages of +11°C in winter, +3°C in summer, and +7°C for the whole year, he says that the subtropical climate zone of the northern hemisphere would be shifted at least eight hundred kilometres north of its present position in winter, but only about two hundred kilometres north of its present position in summer. But at the same time, the climate zones of the southern hemisphere would have changed only a little compared with the present day, in line with the modest global mean increase in temperature of 4°C.

From a combination of calculations of these climatic shifts and the geological record of conditions around the world three million years ago, Flohn concludes that there would be a drastic reduction in the winter rains of California, the Mediterranean, the Middle East and the Punjab region of India. They would lie in a desert belt equivalent to large parts of Mexico and the Sahara today.

228

The hottest region of the Earth now, the 'meteorological equator', already lies about 6° north of the geographical equator because of the uneven geography of the globe, with more land in the northern hemisphere and a bigger icecap in the south than in the north. But when the north polar icecap melts, as it will do before the year 2050 on current projections of the greenhouse effect, the meteorological equator will shift northward to between 10 and 12°N. This takes it up above the latitude of Panama, through northern Nigeria and southern Ethiopia, across the tip of India north of Sri Lanka, and through the Philippines. This would squeeze the belt of summer tropical rains into a band between about 12 and 20°N, extending into the Sahara, Sudan and Mexico. They would all become regions of lush 'tropical' vegetation. But in the south, the same amount of rainfall would be spread over a belt from 12° *north* of the equator to 20°S. Large parts of this climate zone would become desert. This includes Brazil, the tropical heartland of Africa (Gabon, the Congo, Zaire . . .), Malaysia, New Guinea and even northern Australia. The general increase in aridity caused by a lack of rainfall would be exacerbated by increased evaporation from the land in a hotter world. Associated changes in ocean currents, says Flohn, are likely to suppress El Niño entirely, reducing the evaporation of water from the eastern Pacific and making the South American drought even worse.

All of the other changes I have described can only be regarded as milestones along the road of global warming, landmarks that we will whizz past before Gaia has time to adjust to the new and changing 'boundary conditions'. But Flohn's fourth scenario is different. Although concerted action by some nations might slow the warming trend, I can see no realistic prospect of bringing the greenhouse effect under control in time to prevent a global warming of 4°C. But when that happens, and the Arctic icecap melts, the world will change so drastically that either civilization will collapse, halting the build-up of greenhouse gases, or we will come to our senses and stop the global geophysical experiment ourselves. Either way, the scenario of the lopsided Earth represents a genuine *equilibrium* state that the world is likely to settle down to in the twenty-second century. If we are careful, the critical warming that triggers the change may be delayed until the late twenty-first century, and I won't be around to see the Arctic ice disappear. If we are profligate in our use of fossil fuel and compounds such as CFCs, the dramatic changes described in the previous paragraph will happen in my lifetime. As a scientist, I *almost* hope to live to see it; as a human being, I definitely want to see the pace of change slowed. Given time, human society can adapt to almost anything; the great danger of the lopsided climate is that it will be upon us in a rush, once the critical temperature needed to start the feedback that melts the Arctic ice is reached. Whenever it happens, it will be a disaster.

You may think I am being unduly pessimistic in being so certain that it will happen. Surely, you might hope, we can bring greenhouse emissions under control within the next twenty years? Alas, even that is unlikely to save us, or our children, from the lopsided climate. There are other feedbacks likely to be triggered in much less than twenty years that are just waiting to carry us over the critical threshold.

Stings in the tail

One of the additional greenhouse threats comes from a source I have already mentioned – peat. Around the Arctic Ocean, as the forests advance northward the present huge expanses of tundra will shrink almost to nothing. Below the top few centimetres of vegetation in this vast, almost uninhabited region of the globe there lies an enormous depth of permanently frozen peat, which stores about fourteen per cent of the organic carbon on Earth today. As the ice melts, this peat will begin to decompose and will eventually release most of that carbon to the atmosphere in the form of carbon dioxide. The size of the additional greenhouse effect this will produce over the next few decades is impossible to predict accurately. But what matters is that it is a positive feedback in which global warming causes the release of more carbon dioxide to the air and thereby increases the global warming.

A similar process operates in the soil of forest floors, rich in organic carbon. Gundolf Kohlmaier, of Frankfurt University, has been studying the way in which carbon is exchanged between humus in the soil, standing vegetation and carbon dioxide in the air. Other researchers had already drawn attention to the way in which forest clearance allows the humus to oxidize, adding an extra burden of carbon dioxide to the air, over and above the amount produced from burning the carbon in the wood of the trees themselves. Kohlmaier calculates that a temperature increase alone will change the workings of these natural carbon cycles, even where forests are still standing, speeding the decay of organic material in the soil. This would add more carbon dioxide to the atmosphere than in any of the scenarios I have discussed so far, and would strengthen the greenhouse effect still further. Although this is a new study and the detailed figures may be questioned, once again the important point is that this is another positive feedback, enhancing the strength of the greenhouse effect.

The third sting may come from methane. Recent investigations of the material of the continental shelves suggest that there may be huge quantities of methane stored in the form of methane hydrates – perhaps as much as ten thousand billion tonnes of carbon locked up in this form. The methane deposits are buried at depths of several hundred metres, and are locked up with water molecules in a form known as clathrate. The clathrate is kept stable by the pressure of water and sediment pressing down on it, and may contain more carbon than all of the known reserves of coal in the world. Some of this methane – nobody can be sure how much – is likely to be released either if the pressure above falls (which is unlikely) or if the deposits warm up (which is very much on the cards).

As well as being located beneath the sea in continental margins, methane hydrate is stored deep below the permafrost around the Arctic. There is some evidence that hydrates in the Arctic region are now decomposing, releasing methane into the atmosphere. This could explain, rather neatly, the way in which variations in global mean temperature and in the concentration of methane in the atmosphere march almost precisely in step over the length of the Vostok core, a span of 150,000 years. And with methane the second most important natural greenhouse gas, it is, of course, yet another positive feedback.

It is the existence of these natural feedbacks reinforcing the growth of the greenhouse effect initiated by human activities that makes it so difficult to see how the global warming can be halted short of an increase of 4°C over the temperatures that prevailed a hundred years ago. But if there is to be any hope at all of preventing the world from tipping into a disastrously lopsided pattern of climate before the end of the twenty-first century, action must be taken now. There is no time to waste in cleaning the crystal ball to improve our forecasts, since the picture is clear enough to show that disaster lies ahead. In 1978, a group of specialists assembled by IIASA to look into the problem of the anthropogenic greenhouse effect concluded that 'mankind needs and can afford a time window of between five and ten years for vigorous research and planning ... to justify a major change in energy policies'.* I am writing these words in the last month of 1988. The ten years are up, and the time for action is now. So – what can we do?

*Jill Williams, editor, *Carbon Dioxide, Climate and Society*, IIASA Proceedings Series Environment, Volume 1, Oxford University Press, 1978.

TEN

What To Do

The 'worst case' effects of the anthropogenic greenhouse effect – a shift into a lopsided climate with an ice-free Arctic Ocean – will occur when the carbon dioxide equivalent of the atmosphere reaches a concentration of about eight hundred ppm. This will lead, first, to a series of catastrophic weather extremes over a period of perhaps twenty years as the weather machine adjusts to its new boundary conditions. As the weather settles down into a new pattern, climate zones in the northern hemisphere, where by far the bulk of the world's human population resides, will be six hundred kilometres (or more) north of their present positions. Such a prospect seems unacceptable to many people – but it will require a major change in our global way of life if it is to be avoided.

Nobody can predict exactly what will happen to the climate in the twenty-first century, nor how society will respond to the growing awareness of the threat. But change, in both climate and society, is inevitable. It is too late for even the most draconian measures to prevent a further rise in global mean temperatures of at least 1°C, and part of any long-term planning for the future should take account of this. One response to the growing strength of the greenhouse effect is adaptation – which might mean anything from an individual decision to stop taking summer holidays in places like Spain or Greece to a farming community's decision to shift from wheat to maize, or a government's plan to build new reservoirs. But if the increase in the greenhouse effect cannot be stopped, it can certainly be slowed. The second response to the perception of the reality of the greenhouse threat should be action at all levels, from individuals to international agreements, to keep the pace of change down to something that we might have a chance of adapting to.

233

Neither you nor I can wave a magic wand and ensure that appropriate action to minimize the build-up of greenhouse gases is taken – in the real world, decisions on the construction of power stations and transport networks, for example, are made on political grounds, taking into consideration a variety of factors, including the politicians' wish to be re-elected. But I can tell you what sort of decisions ought to be taken if the greenhouse threat is, as I believe it to be, of overriding importance. If enough people feel the same way, and make their views known, then the politicians' wish for re-election may itself ensure that they take the greenhouse threat seriously.

One of the most interesting features of any discussion of policies to minimize the global rise in temperatures is that in almost every case these are policies that are desirable on other grounds as well. The problem of CFCs provides a good example. These gases entered the arena of political debate because of the damage they are doing to the ozone layer of the stratosphere (see Appendix). But as we have seen they are also powerfully effective greenhouse gases, making a major contribution to the shift of the Earth into a hothouse state. Agreements aimed at limiting damage to the ozone layer so far talk only of reducing the rate of release of CFCs to about half of the 1988 level, but most scientists who have studied the problem are now calling for a complete ban on the use of these gases, except in a very limited number of 'essential' applications, notably medical. Such a total (*and immediate*) ban on CFC emissions would not solve the greenhouse problem, but could, on its own, give us a breathing space of perhaps ten years before the date when the effective carbon dioxide concentration of the atmosphere reaches twice the pre-industrial level. Ten years is not long, but in the situation we now face every little helps. There are those, indeed, who think that this is more important than limiting the damage to the ozone layer. Jim Lovelock, who invented the CFC sniffer and was the first person to use it to trace the spread of CFCs around the globe, once said, in a paper in *Nature* in 1973, that the presence of these gases in the atmosphere 'constitutes no conceivable hazard'. At an international meeting on ozone depletion held in London at the end of November, 1988, he ruefully reminded the audience of this remark, and said that he had since revised his opinion. He now sides, he said, with those who feel that an immediate and total ban on all emissions of CFCs is necessary – because of the implications for the greenhouse effect. Of course, whatever your motive for pressing for such controls on CFC emissions, there will

234

be an additional benefit in the form of a reduction in the damage being caused to the ozone shield. There is, though, another side to the coin. Those who are pressing for CFC controls on the grounds of the damage being done to the ozone layer must take care to ensure that any substitutes developed for these CFCs are not only 'ozone friendly' but are also 'climate friendly' and do not contribute to the greenhouse effect.

Controlling CFC emissions is easy, in the sense that we can imagine a world without the convenience that these gases provide. But we cannot imagine our society adapting to a world without power stations, steel mills, high speed transport and intensive agriculture. That is why the temperatures in the global hothouse will continue to rise, and why even slowing the rate of that rise will be difficult and painful – but not as painful, ultimately, as the consequences of letting the temperature rise proceed unchecked.

Burn less

The most direct way to reduce emissions of greenhouse gases (after a ban on CFCs) is to burn less fossil fuel – more realistically, we should aim to minimize the increase in the rate at which such fuels are being consumed. There is no way in which we can stop adding carbon dioxide to the atmosphere, but many studies have shown that it is possible to reduce the rate at which the carbon dioxide concentration is building up, even while maintaining economic growth and coping with an increasing human population (but remember that carbon dioxide contributes only half of the greenhouse problem). Ninety per cent of the commercial production of energy in the world today is from fossil fuel, but a great deal of the energy that is produced is wasted – a hangover from the days of cheap energy. Just how wasteful we used to be is reflected in the way the growth in world energy consumption declined from a rate of nearly five per cent a year in the early 1970s to two per cent a year now, following oil price rises and economic recession. Part of the reason for this decline is that some consumers use energy more efficiently simply because it costs more; on that basis, energy is still far too cheap, and a suitable tax might focus the minds of individuals and industries even more firmly on how to save energy (with the added bonus that revenues from the tax could be used for research into ways to alleviate the impact of the greenhouse effect). Many of the appropriate measures to save energy are familiar from the long debate about the way we ought to live. Better insulation in houses and other buildings (it doesn't cost any more to build a properly insulated house than to build one which leaks heat); cars that use fuel more efficiently (prototypes already exist that will run for ninety-eight miles on a gallon of fuel, although most cars in the US manage less than a quarter of this figure); improving the efficiency with which energy gets from the power station to the consumer (in the UK, more than sixty per

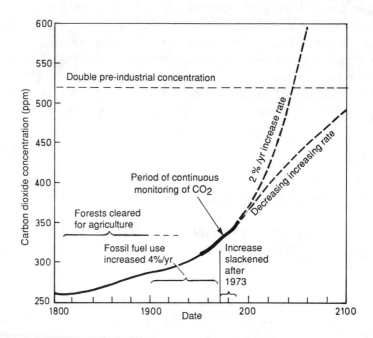

Figure 10.1
Past changes in the concentration of carbon dioxide in the atmosphere (based on a slightly low estimate of the pre-industrial concentration of 260 ppm), and two projections into the future. Continuing growth in emissions at a rate of two per cent a year probably represents the most rapid build-up of carbon dioxide we are likely to see: the 'decreasing increase rate' line is a slightly optimistic projection based on a slow-down in the use of fossil fuel that would bring emission rates back to the level of the 1980s over a period of fifty years. The actual future trend is likely to lie between these two lines, bringing a doubling of the 'natural' concentration of carbon dioxide alone some time between 2030 and 2100. This will be enough to raise the equilibrium mean temperature of the Earth by 3°C; other anthropogenic greenhouse gases will be contributing an additional warming roughly as big as the carbon dioxide greenhouse effect.
(Source: Will Kellogg, NCAR)

cent of the energy available in coal, oil or gas literally goes up in smoke and never reaches the user).*

Some of these changes require action at government level, but even individuals can make their own contribution. All those things we were encouraged to do the last time an 'energy crisis' hit – sharing cars, riding bicycles, turning the central heating down a couple of degrees and wearing a sweater – are still relevant. And there is new technology available to make you feel less guilty about using electricity. Compact fluorescent light bulbs that provide the same light as conventional incandescent bulbs but use only a quarter of the electricity are in the shops (some shops) today. At present, these bulbs cost more than conventional bulbs, but last more than five times as long. Replacing a single seventy-five-watt incandescent bulb by a fluorescent bulb that uses only eighteen watts of electricity but is just as bright eliminates the burning of 180 kilogrammes of coal over the lifetime of the bulb, and saves the consumer money in the form of a smaller electricity bill. If electricity prices were higher, the incentive to use these bulbs would be greater, their cost would come down as more were sold, and the value of the new bulbs to the household would be greater still – a beneficial example of positive feedback.

Even before this invention became available, a study carried out by researchers at Harvard University at the end of the 1970s concluded that the energy consumption of the US could be reduced by thirty to forty per cent with 'virtually no penalty for the way Americans live'.† Advocates of the 'green' movement would like to see humankind living more in tune with the environment and with other species of life on Earth for many reasons, but the problem of the greenhouse effect is certainly one of them. Studies such as Amory Lovins's classic *Soft Energy Paths* show how virtually all of the energy requirements of the developed world could be met

*This isn't due to incompetence, but a result of the basic physics of the way heat is generated and converted into electricity. One way to get more useful energy out of the same amount of coal or oil is to build small power stations close to the communities they serve, so that the heat that would otherwise be wasted as so much hot air can be piped to the buildings people live and work in to heat radiators or water. Such systems are known as combined heat and power, or CHP, installations, and we need a lot more of them, instead of huge, remote electricity generating stations, to help slow the growth of the greenhouse effect.

†R. Stobaugh and D. Yergin, editors, *Energy Future* (Random House, New York, 1979).

in the mid-twenty-first century by a combination of environment friendly systems, such as solar energy, wind and wave power, geothermal heat and controlled use of biomass, including wood. Such an extreme view of what some see as an ideal world would involve dramatic changes in society, and has little or no prospect of becoming reality. Nevertheless, many of the possibilities discussed by the idealists could be incorporated into society to some extent without major disruption; and, once again, every little helps.

Overall, in the industrialized world the present consumption of energy per head could be cut in half by the year 2050 with the aid of better insulation in buildings, refrigerators (already designed) that use thirty per cent less electricity, cars that use a third as much fuel as the present generation, and so on and so on. Look at it this way. Demand for energy is scarcely growing at all in the rich world, so we need only make our use of energy more efficient by one per cent each year to cut demand in half, in the developed world, by the middle of the next century. Just which technologies or economies are used to achieve this scarcely matters as long as the job gets done.

But what of the Third World? It is all very well arguing for more efficient use of energy in countries such as the US and the members of the European Community, where a little belt-tightening wouldn't cause much pain. But if the poor regions of the world are to achieve a satisfactory standard of living they will need to increase their use of energy. The best any optimist might hope for is that this increase might be countered by improved efficiency in the rich world, leaving annual emissions of carbon dioxide constant at present levels, steadily strengthening the greenhouse effect; but this is not very likely. In one sense, though, the problem faced by developing countries is more straightforward than that faced by the developed world. Rich countries have to undo some of the mistakes made in the past. But countries that lack an infrastructure of existing inefficient energy production and use are well placed to build efficient systems from the start – provided they have the money to pay for them. Once again, the correct response to the greenhouse problem strikes at the heart of another major global issue. Third World countries are poor, in large measure, because they are servicing crippling interest charges on their debts to the rich north. This is immoral and inexcusable on many grounds; wiping out those interest charges, or writing off the debts themselves, could go a long way towards providing the funds for investment in appropriate energy technologies. Indeed, provision of such energy policies could be a condition of the agreement under which the debts were eased. Rich countries might feel, at first sight, that easing the financial burden on the Third World might damage their own economies; but apart from the moral issue this might well be cheaper, in the long run, than waiting for increasing temperatures and rising sea levels to do their own damage to the economies of countries such as the US and Britain.

Not that there need be much philanthropy in any of this. Investing in more efficient sources of energy is an attractive prospect for even the most ardent free-market capitalist. One study of energy usage in Brazil, for example, showed that investment in more efficient refrigerators, street lighting and lighting in buildings, at a cost of about ten billion dollars, would save forty-four billion dollars between 1985 and 2000. The investment has not been made, but the example is still valid. At an even more fundamental level, one of the major problems in the poorest parts of the world is obtaining wood for cooking fires. As populations have increased, forests have been stripped, contributing to the greenhouse effect and causing erosion of valuable land. Closed cooking stoves require only half the fuel of traditional open fires, and it has been estimated that it would cost about one billion dollars a year to provide efficient stoves to all rural households in the Third World. This is a tiny proportion of the Gross Global Product, but would provide benefits to millions of people and to the environment.*

Change the mix

Still on the subject of fossil fuel, in the short term we could gain some breathing space by changing the mixture of fuels being burnt. Coal produces more carbon dioxide than oil, and oil much more than natural gas, if each type of fuel is used to provide the same amount of energy. Synthetic fuel oils derived from coal – substitutes for oil and gasolene – produce even more carbon dioxide than coal itself, which is the main reason for the opposition among environmentalists to the US 'synfuels' programme of the 1970s. Switching the source of energy for power stations from coal to natural gas would be a help in slowing the growth of the greenhouse effect. Unfortunately, sixty-five per cent of total energy

*Both these examples are taken from the Friends of the Earth report *The Heat Trap*, written by J. W. Karas and P. M. Kelly (FoE, London, 1988).

production already comes from oil and gas, and these two sources only account for sixteen per cent and ten per cent, respectively, of known fossil fuel reserves. Even in the medium term, let alone the long term, their contribution to the fuel mix is likely to diminish, not increase. But even though the scope for changing the mix of fuels in use today is limited and short term, providing an option for the next two or three decades only, that is no reason to ignore it. In fact, natural gas associated with oil fields is still very often thrown away – either allowed to escape into the atmosphere, or 'flared off', burning at the drilling site itself and contributing to the greenhouse effect without ever being used to generate useful energy. It simply is not cost effective for all of the gas to be husbanded and transported to places that need energy. Once again, the way to make this cost-effective might be through a system of differential taxation, which hits hardest at fuels that produce more carbon dioxide (as against the present policy, in many parts of the world, of *subsidizing* coal-based energy generation in order to make it competitive with oil). This, however, would exacerbate another potential problem: it is possible to argue that reserves of oil should be saved for other uses, not burnt as fuel. Oil provides the raw material for a huge chemical industry producing, among other things, the plastics that have transformed our lives over the past half century. To take just one example: without plastics to use as insulators our entire global communications network would break down (TV sets built in the 1930s used *paper* to insulate their wiring; under modern safety regulations, such lethal objects would never be allowed in the home). Our descendants may well curse the wastrels of the twentieth century who squandered such a valuable resource as oil by burning it.

So, once again, alternative sources of energy come into the picture. They can be made to work, but it won't necessarily be easy. One criticism of systems such as solar energy or wave power schemes is that they produce energy all right, but out at sea or in the desert, not in the places where it is needed. There is an answer to this criticism, but it still needs careful study. Electricity produced from wave power, for example, could be used to crack water into its component parts, hydrogen and oxygen. The hydrogen could be stored under pressure, transported around the world and used as a fuel, while the oxygen is released into the air. When the hydrogen burns, it combines with oxygen to make water again. It sounds very attractive, clean and simple. Indeed, the Soviet Union is developing an aircraft which uses liquid hydrogen as its fuel; in West Germany several car manufacturers, including BMW, have plans to develop internal combustion engines that run on hydrogen. The note of caution comes in because we cannot yet be sure that emitting large quantities of water vapour into the atmosphere will be a thoroughly good thing. With any luck, the water from car exhausts would condense and run away back into the sea pretty quickly. But what if the world's fleet of aircraft switched to hydrogen fuel? Remember that water vapour is itself a greenhouse gas, so it might make the world warmer. On the other hand, emitting large amounts of water high in the atmosphere might encourage clouds to form, reflecting away incoming solar heat and cooling the globe. As with all technological fixes, there is a danger that the cure will be at least as bad as the disease. And that certainly applies to the biggest and most widely touted technofix for the greenhouse effect.

What not to do

The naive 'solution' to the problem of a build-up of carbon dioxide in the atmosphere is to suggest a switch to nuclear power. Nuclear power stations emit no carbon dioxide, and do not cause acid rain. They are, in a sense, environmentally clean. Even Britain's Prime Minister, Margaret Thatcher, made a comment, in an interview in *The Times* in 1988, to the effect that nuclear power is 'greener' than coal. But a close look at the implications of going for an all-nuclear programme of power generations shows that things are not as simple as they seem at first sight.

The first important point is that any attempt to make nuclear power replace a significant portion of the existing world energy generation capacity would itself consume a great deal of energy. It isn't just the nuclear power plants themselves that have to be built, but the industrial infrastructure required to build the power stations and to ensure that they are kept supplied with fuel. While the world was in the process of switching to nuclear energy generation, it would be using up fossil fuel far more rapidly than it does today because of all this industrial activity. The so-called 'energy payback' time for a nuclear power station is very long; it doesn't produce as much energy as was used in building it for many years. The building programme itself would take decades, and the benefits in terms of carbon dioxide concentration of the atmosphere would not be felt for further decades, by which time the Arctic ice would probably have melted. So a rush programme of building new nuclear power stations and scrapping existing coal-fired power stations would do more harm than good, simply in terms of the amount of carbon dioxide released over the next few crucial decades.

Nor would such investment be cost effective in other terms. Bill Keepin and Gregory Kats, of the Rocky Mountain Institute in Snowmass, Colorado, carried out a major investigation of the problem, which they published in 1988 and which provided the basis for testimony by Keepin to the House of Representatives on 29 June that year. Leaving aside the question of how much energy is used (and carbon dioxide released) in the building of a nuclear plant, they pointed out that the cost of energy generation from new US nuclear plants is 13.5 cents per kilowatt hour, in 1987 dollars. But the cost of saving electricity through improved efficiency (measures such as insulation of buildings) is only two cents per kilowatt hour. *Even if the nuclear power plants already existed* and were waiting to come on stream, if you have a limited amount of money to spend and the choice is between buying nuclear electricity to replace coal-generated electricity, or making investments in energy conservation so that you use less electricity, conservation will be nearly seven times as effective in reducing the amount of coal being burnt.* Looking at it from a slightly different angle, for every hundred dollars invested in new nuclear power, about one tonne of additional carbon is added to the air as carbon dioxide, compared with the amount that would have been released if the same money had been spent on improving efficiency of energy use.

Nevertheless, Keepin and Kats developed a scenario to give some idea of the amount of work involved in 'solving' the greenhouse problem with the aid of nuclear power. Again, they ignored the input of carbon dioxide into the atmosphere caused by the construction itself, and they made deliberately optimistic assumptions about the cost of each power station and the time it would take to build, set at six years instead of the ten to twelve years that has been experienced recently in the US. Allowing for the need both to replace existing coal-fired power stations and to

---

*The cost of the coal-fired electricity doesn't change the calculation significantly. The 'marginal' cost of the nuclear electricity is the amount by which the nuclear cost exceeds the coal-fired cost, and this makes the ratio smaller. But with conservation, you save the money that would have been spent on electricity, which makes the ratio bigger again. The Snowmass study takes account of all this.

cope with increasing energy demand worldwide, they found that a new thousand-megawatt nuclear plant would have to be completed roughly every two days for the next forty years. A great deal of this nuclear activity would be going on in the Third World; it would cost some six hundred billion dollars a year; and even while it was going on carbon dioxide from plants fuelled by oil and natural gas would still be adding to the burden of greenhouse gases in the atmosphere.

Quoting a report in the April 1988 issue of *Scientific American*, Keepin told the House that, by contrast, if all new buildings were designed to be energy efficient, the US alone would save the equivalent of eighty-five such power plants and two Alaskan oil pipelines in fifty years' time. More efficient energy use would save consumers $110 billion a year, but would cost no more than fifty billion dollars a year. 'It's as if,' he said, 'we were offered sixty billion dollars a year to live in a cleaner environment.' And that is by no means the limit of potential savings. Raising the average fuel efficiency of US cars from about eighteen miles per gallon (the present figure) to just twenty-eight mpg (still below the average in the European Community today) would save the equivalent of *all* OPEC imports into the US.

As long as new power stations continue to be built, there will always be a case, on specific occasions, that nuclear power may be the best option, and in that way nuclear power can indeed (and does already) do something to reduce the amount of carbon dioxide getting into the air. But it is by no means 'the' solution. Nor is the other bright idea of the techno fans, that of taking carbon dioxide out of the factory and power station chimneys and getting rid of it underground or in the sea.

Once again, the idea is attractive at first sight, even if you 'buy' the nuclear option. After all, nuclear power stations cannot displace a very large proportion of greenhouse gas emissions. Only half of the present greenhouse emissions are carbon dioxide, and fossil fuel-fired power stations contribute only fifteen per cent of all carbon dioxide emissions. Extracting carbon dioxide from stack gases could be applied not only to power stations but to factory smokestacks as well – if only it were not prohibitively expensive in terms of both money and, you guessed, energy.

At least fifty per cent of the carbon dioxide in stack gases can be recovered, in principle. But if we consider the specific example of a coal-fired power station, the energy needed to extract something over half the carbon dioxide would be forty per cent of the energy produced by burning the coal in the first place (and a comparable amount of energy would be needed to remove carbon dioxide from the gases emitted by, say, a steel mill). The gases have to be captured (probably by electrically powered fan systems), compressed, cooled, perhaps treated chemically and stored at low temperatures, all of which takes energy. If it takes nearly half the energy of the power station to remove just over half of the carbon dioxide, you have to build another power station the same size to make up for the shortfall in energy, and you end up with very nearly as much carbon dioxide going into the air while the cost of electricity has almost doubled – the economics of the madhouse.*

There is also the problem of what to do with all the carbon dioxide once it has been captured. The most widely touted bright idea is to inject it into the oceans, either as a cold, dense liquid or as a solid, so that it would sink into the deep currents and be removed from contact with the atmosphere for hundreds or thousands of years. Apart from the fact that shipping the carbon dioxide off to a suitable site would use up yet more energy, there are two strikes against this idea. The first is that we do not know how such a major change in the amount of carbon dioxide dissolved in sea water will affect natural carbon cycles. The second is that this is no solution to the problem at all, but at best postpones the day of reckoning for a few centuries, leaving our descendants to cope with our mess when the deep currents return to the surface and the stored carbon dioxide comes into contact with the atmosphere.

Probably the maddest of all such proposals, though, is the idea that carbon dioxide from the emissions of fossil fuel-fired power plants could be turned back into useful fuel using energy from solar or nuclear power stations. This breathtakingly bizarre scheme would involve the construction of about six thousand plants, each using a thousand megawatts of energy. As well as all the problems of the carbon dioxide released during their construction, this additional number of large energy sources on the surface of the Earth would produce so much waste heat that they could change the climate directly, without any help from the greenhouse effect.

*Activists in the 'green' movement also face a conflict of interests in trying to clean up other emissions from fossil fuel-fired power plants, such as the sulphur dioxide that causes acid rain. Flue gas desulphurization at a major power station takes energy, and in effect reduces the efficiency of the power plant by about five per cent. This means that about five per cent *more* carbon dioxide gets into the air than from a comparable power station without desulphurization, while the cost of electricity generated by the station increases by about ten per cent because only about half of the energy produced gets to the consumer in a useful form.

I have, perhaps, travelled too far into the realms of science fiction. But there are less crazy ways in which carbon dioxide might be taken out of the air, using far less technology and working with the natural cycles of Gaia herself.

The return of Johnny Appleseed?

If the technology of removing carbon dioxide from the air is impractical, can biology help? Trees and other plants are largely made of carbon, obtained from the carbon dioxide in the air, and some trees have very long lifetimes. The greenhouse effect has certainly been made stronger in the past by destruction of forests, and it would be very pleasing if we could alleviate the greenhouse problem, if only in the relatively short term, by planting new forests and letting them grow. It sounds like a hippy fantasy; but for once the appealing 'natural' option does have something more going for it than wishful thinking. Reforestation is being taken seriously as a policy option, not only because every little helps in our efforts to slow the growth of the greenhouse effect, but also because forests slow erosion, increase the availability of water, provide some timber even if the bulk of the trees are being left to grow, and provide a home for the diversity of life that is in danger in many parts of the world today. Even if efforts at reforesting large parts of the world do not turn out to have much impact on the build-up of carbon dioxide, no harm will have been done by planting the trees, and a great deal of benefit will have accrued in other ways – so why not give it a go?

Gregg Marland, of Oak Ridge National Laboratory, in Tennessee, has carried out a major study of what might be involved in any serious attempt to control the greenhouse effect by planting trees, and he testified on the subject to the Senate energy committee in Washington in 1988. The amount of carbon getting into the atmosphere as carbon dioxide from the burning of fossil fuels is about five billion tonnes each year. To provide a rough guide to what might be achieved by biological fixation of atmospheric carbon, Marland calculated how many trees would be needed to convert this much carbon into wood each year. He based his calculations on the American sycamore, a species that has been well studied and which has a good carbon uptake rate. Other species growing in the tropics would probably be even more effective at absorbing carbon dioxide, but on a tree plantation in Georgia one hectare of sycamores can absorb 7.5 tonnes of carbon every year as the trees grow. To fix five billion tonnes of carbon a year would need seven million square kilometres of young trees, an area about the size of Australia. As it happens, this is also about the area of all the tropical forest cleared as a result of human activities since the end of the latest ice age, most of it during the past few hundred years.

A forest the size of Australia does seem rather large. But the trees need not all be in one place, and nobody would seriously try to absorb *all* of the carbon dioxide being emitted by power stations, transport and industry each year. Half the carbon dioxide being produced is already being absorbed by natural sinks,* so a forest just half as big might be enough to halt the build-up of this particular greenhouse gas, while one even a quarter the size would make a valuable contribution to reducing the growth of the greenhouse effect. An area roughly equal to the land mass of Zaire would be ample to do the job.†

Of course, even if it could be achieved, storing carbon in trees is only a temporary solution to the problem. Eventually the trees mature, and sooner or later the carbon they store will be released back into the atmosphere as they die. Some of the trees might be used as fuel for power stations, instead of coal, while new trees are planted to replace them; some of the mature trees might be cut down and stored underground (in abandoned coal mines?) But maybe these are worries that we could reasonably leave to the next generation; if we can slow the build-up of greenhouse gases enough to give society time to adapt to the changing climate and to reduce its dependence on fossil fuels, deciding what to do with the new forests when they mature in the twenty-first or twenty-second centuries shouldn't be an insuperable problem.

Although it might seem foolish to begin replanting at a time when many tropical forests, in particular, are still being destroyed,

---

*But every time I see that statistic I worry that those natural sinks may be getting overloaded and that some day soon they may stop taking up so much carbon dioxide, leaving us with an even bigger problem on our hands.

†A related idea, which might be easier in practice, is to fertilize the oceans by spreading nitrogen and phosphorus compounds (and perhaps iron compounds) on the surface to encourage the growth of phytoplankton. But this is more likely to have unpredictable and perhaps unwelcome repercussions on the natural carbon cycles, whereas planting trees would, in a sense, simply be restoring the status quo that existed before human intervention.

reforestation also has the advantage that it provides a way for anybody to do something useful to help. In October 1988, the American power company Applied Energy Services, of Arlington, Virginia, agreed to help offset the production of carbon dioxide from a new power station in the US by planting fifty-two million trees in Guatemala. The scheme – the first of its kind in the world – will operate through CARE, the international relief and development agency, which is being given two million dollars by AES to help forty thousand farmers to plant the trees during the 1990s, as part of a land development programme. A further sixteen million dollars for the project comes from American government aid agencies. The power station will emit 387,000 tonnes of carbon dioxide each year during its forty-year life, and the growing trees will absorb at least that much carbon dioxide, as well as providing other benefits to the Guatemalan farmers.

Saving existing tropical forests from destruction would be just as beneficial, in terms of the greenhouse effect. This would also be beneficial for many other reasons, often aired in debate about the damage we are doing to our planet. In one season in 1988 an area of rainforest the size of Scotland was burned in Brazil to provide pasture land for cattle. The cattle, in many cases, provide meat for North America hamburgers; the pasture lasts only a few years before the soil is worked out and destroyed, then another huge area of forest is burned (in fact, this much forest is cleared *every year* in a continuous cycle). The problem is inextricably linked with the crisis of development facing the poor regions of the world. Brazil is one of those countries with a huge foreign debt, which must be serviced by earning foreign currency. If the way to get that currency is to burn the forest and rear hamburger meat, can the Brazilians really take sole blame for the damage that is done to the planet – Gaia – as a whole? Global problems need global solutions, and if the rich countries of the world are worried about rising sea levels and other consequences of the greenhouse effect, why shouldn't they offer simply to buy the forest for the cost of the Brazilian debt or even for more, on the single condition that it is left alone? At worst, that might mean a temporary and minor tax hike for citizens of the rich world; I'll gladly pay my share, and do without hamburgers. Or do the people of North America actually value their cheap burgers more than they value the stability of the climate? The 'solution' may be naive and simplistic; but at least it indicates the magnitude of the changes required even to bring the build-up of carbon dioxide under some sort of control. And that still leaves other greenhouse gases to worry about.

Good husbandry

Although CFC production could conceivably be drastically reduced, if not banned entirely, and the growth in emissions of
carbon dioxide might be slowed, it is very hard to see how the
contribution of other greenhouse gases to the global warming can
be prevented from increasing rapidly. The build-up of methane
is very closely linked with increasing agricultural activity, and as
the population of the world continues to grow agricultural activity
must intensify in order to feed the population. Of course, as some
people like to point out, all these problems would disappear if the
human population ceased growing, but that doesn't really seem a
very helpful comment in the real world. Both ozone and nitrous
oxide are released into the lower atmosphere as a result of the
combustion of fossil fuel, either from vehicle exhausts or from
power stations and factory smokestacks. Any measures that reduce
the rate at which the build-up of carbon dioxide is increasing will
help to control these two greenhouse gases as well. But nitrous
oxide is also produced from nitrogen based fertilizers, and once
again it is hard to see how the use of these fertilizers can be
reduced in a world where there are more mouths to feed each
year.

In fact, the fertilizer issue is not at all straightforward. In
some countries, including parts of the US and Europe, there is
a view in some quarters that land is already being over fertilized,
with detrimental effects on the environment including pollution of
drinking water. A reduction in fertilizer use in the rich world might
make those countries better places to live, as well as providing more
fertilizer for hungrier regions. Then again, some researchers argue
that massive application of fertilizer and other kinds of agricultural
technology imported from the rich north is not an appropriate
response to the problem of feeding people in the Third World
anyway. Instead of copying a farming revolution that was developed
for temperate latitudes, tropical and sub-tropical regions might be
better off developing their own kind of good husbandry based upon
the traditional practices of the peasant farmers who were more in
tune with the land – just as the northern agricultural revolution
developed over hundreds, even thousands, of years. Whatever the
outcome of that debate, though, the impact on the greenhouse
effect is likely to be small compared with the contributions from
methane, carbon dioxide and (I fear) CFCs in the first half of
the twenty-first century. Nothing is going to stop the greenhouse

effect from intensifying; so how might we begin to adapt to the changes ahead?*

Adapt – or die?

If greenhouse gas emissions continue to rise unchecked, the rate at which temperatures will increase over the next half century or so will be at least 0.3°C per decade. If an intensive effort is made to bring those emissions under control, this might be slowed to 'only' 0.1°C per decade.† That would be about the same as the maximum rate of change of climate systems before human intervention, and should be something that both human society and the world at large could cope with; but unless wholehearted and worldwide efforts are made to slow the change, it is likely to proceed at least twice as fast as anything the world has seen before. As Gus Speth, President of the World Resources Institute, said in 1986, 'Inaction is not a neutral, low risk policy, but rather a gamble that risks much greater harm.'

Clearly, we must try to adapt to the coming change. But the first problem is that we cannot be sure exactly what that change will involve, in detail. Power stations, sea defences and transportation systems need to be planned on a timescale of decades, but we do not know what kind of local climate to plan for in, say, Milwaukee in 2001. Depending on what decisions governments make *now* about controlling the problem, the global mean rise in temperature between now and then might be 0.1°C, 0.3°C, or even 0.5°C; and nobody can yet say just what any of these changes will mean for specific localities. Even governments that have grasped the importance of the greenhouse effect still do not seem to have fully appreciated the need for much more scientific research into the problem in order to make planning and adaptation easier. In Britain, for example, the Antarctic Survey is at present one of the favoured research institutions receiving a major cash

*If you want to look into the agricultural arguments, I spelled them out in my book *Future Worlds* (Plenum, 1981), describing the work of the Science Policy Research Unit at Sussex University.

†Figures from *Developing Policies for Responding to Climatic Change*, Jill Jaeger, WMO/UNEP report TD-No.225, 1988. On the basis of the evidence summarized earlier in this book, I would set the upper figure at least as high as 0.5°C per decade.

boost. This is largely because it was members of that group that first identified the hole in the ozone layer over Antarctica. The discovery was part of a routine monitoring programme that had been running on a shoestring budget since 1957, the International Geophysical Year. Why? Just for the hell of it – because scientists like to know what is going on. Exactly the same motivation, and a comparably small budget, led to the beginning of the carbon dioxide monitoring programme on Hawaii the same year. Both projects have come close to being cut, on financial grounds, in the intervening decades. If they had been, we might still be unaware of the gravity of the greenhouse problem, or of the existence of a hole in the ozone layer. Yet the same government that now holds up the Antarctic Survey as a shining example of British science, and provides funds to upgrade the research station at Halley Bay in Antarctica, has recently kept the purse strings for other areas of science so tight that another routine monitoring programme is, at the time of writing, in danger of closure. Like monitoring stratospheric ozone or atmospheric carbon dioxide, this began, decades ago, as the kind of long-term project that scientists do just for the hell of it. In this case, part of the project involves monitoring the growth of plankton in the North Atlantic ocean and the North Sea. Plankton are part of the natural carbon cycles; understanding how their growth varies from year to year could provide key insights into the build-up of carbon dioxide in the atmosphere. But because the project has not yet produced any spectacular results, it is under threat.

Still at sea, scientists from forty countries met in Paris in 1988 to initiate the World Ocean Circulation Experiment (WOCE). This is a huge, five-year project to study the role of the oceans in determining climate. It is of key importance in improving greenhouse effect forecasts, and it should cost about six hundred million dollars – chickenfeed, compared with the cost of developing a new military jet fighter or building a few new motorways. But at the time of the Paris meeting, so few funding agencies had confirmed their support for the project that it seemed likely that it might have to be cut by a third, abandoning plans for a detailed study of the circulation system (gyre) of the North Atlantic.

The scientific will is certainly there – and the expertise. In 1986, the International Council of Scientific Unions (ICSU) launched the International Geosphere-Biosphere Programme (IGBP), hailed as the successor to the International Geophysical Year of three decades before. Its aim is to foster studies of the whole Earth system, including interactions between the oceans and the atmosphere, and the workings of the thin green smear on the surface of the planet, the biosphere. Even if ICSU succeeds in cajoling funds out of governments and agencies, it will take ten years for the IGBP to come up with a new understanding of what is happening to our planet. But the first step in any attempt to adapt our lives to the changing climate must be to put such research programmes on a secure footing, to provide the information that planners and policy makers need. Otherwise we are simply shooting in the dark. It seems to me that the IGBP should be given the kind of support that went into the Manhattan Project to develop the atom bomb in World War Two – but on a fully international basis, and without the secrecy.

This isn't just special pleading by an interested party. My own scientific background is in astronomy, and when it comes to funding the search for scientific truth I'd always make a special plea for investigations of the cosmos to come high on the list. But we aren't talking about the abstract search for scientific knowledge now, even though that is how projects like the carbon dioxide monitoring programme began. We are literally talking about life and death for millions of people. The cosmos will still be there for human scientists to study at leisure, if and when we overcome the crisis of the global hothouse. Meanwhile, if funds for essential research like WOCE can only be found by cutting back investigations of black holes and ceasing to probe the mysteries of subatomic particles, then so be it.

While investigations such as the IGBP are going on, there are some actions that can be taken, and will be beneficial, whatever happens to regional weather patterns as the world warms. The initial aim should be for flexibility – a shift away from the monolithic systems that have been a feature of recent history. For example, plans for reservoirs, or sea defences like the Thames Barrier, have been based traditionally on an assessment of past patterns of wind and weather – either the 'climatic normal' of 1950 to 1970, or the past few hundred years. Yet we know for sure that these weather patterns will not be repeated over the next hundred years. Future plans should allow for the possibility of more extreme variations. If a nuclear power station is considered essential, then it should not be built on the coast at sea level, but on higher ground – just in case the sea rises. There should be more effort devoted to breeding types of crop plants that can withstand spells of either more or less rain, higher or lower temperatures, than is ideal for them, and still produce a decent crop – rather than the approach of the past few decades, in which crops were tailored to produce huge yields under ideal conditions, but all too often produced little or nothing if conditions were less than ideal.

Here's one specific example. In the US and Europe, wheat is a staple crop today. The kind of climatic shift that is anticipated for the early part of the twenty-first century would suggest, at first sight, a shift to crops such as sorghum or rye. In fact, in many parts of the world where such crops *ought* to be the dominant grain today, there is a shift away from those traditional crops and towards wheat, because wheat is perceived as being the rich person's crop. A poor country that grows wheat instead of sorghum in some way feels that it has moved up in the global hierarchy, an image reflected in the old name for sorghum in the days of the British Raj, 'kaffir corn'. People don't like to change their diet without some incentive. The shift from sorghum to wheat is driven by the wish to appear successful and rich, either as an individual or a nation; shifting the other way would be difficult, because it would be seen as a retrograde step (so maybe one useful way of adapting to the greenhouse effect might be for governments in the rich world to sponsor an advertising campaign promoting sorghum-based foods as desirable and superior to those based on wheat). But in any case, as the world warms conditions in, say, Britain will not be exactly the same as those further south today. The number of sunlit hours in the day and the pattern of seasons will be different, and even crops that thrive in the heat might need to be adapted to cope with the seasonal variations. Plant breeders usually take ten to twelve years to come up with viable new varieties, so it may already be none too soon to start breeding strains of sorghum that can cope with short winter days and long summer days.* Genetic engineering might speed the process up, but then society would have to overcome its squeamishness about releasing genetically tailored organisms to grow in the fields and to be consumed by human beings, a prospect some people find as abhorrent as that of a shift to nuclear power.

If I could tell you exactly how to prepare for the climate change ahead, we wouldn't need WOCE or the IGBP (and I would

---

*There's also the problem of supplying all the seed that would be needed if farmers in England, say, decided to switch to sorghum in a big way. Maybe it would be prudent to begin immediately to build up a stockpile of seeds of crops that tolerate heat best.

probably have a Nobel Prize). But I can remind you of some of the changes we can expect, from which you can, perhaps, make your own personal plans to cope, as well as urging your political representative to do something at national level.

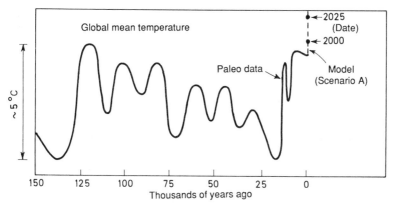

Figure 10.2
The greenhouse effect caused by human activities can be put in perspective against the range of natural fluctuations in the Earth's temperature over the past 150,000 years. Geological evidence ('paleo data') shows that temperatures have varied over a range of no more than 5°C, and that before human activities began to influence climate we were already at the top end of that range. The warming of hothouse Earth will take us outside that range by the end of this century, with temperatures continuing to increase more rapidly than at any time for at least 150,000 years and probably more rapidly than at any time in the history of our planet. This particular projection is from the GISS team's Scenario A; but on this scale all the hothouse projections look much the same.
(Source: James Hansen/GISS)

- By the year 2000, temperatures will be at least 1°C higher than before the build-up of greenhouse gases began. There will be serious droughts in California, Spain and the near and Middle East, and in Australia and New Zealand.

- During the 1990s, because of changes in the atmospheric circulation pattern, the temperate latitudes of the northern hemisphere, including the US, Canada and Europe, can expect some severe winter blizzards, in spite of the rise in average annual temperatures.

- Hurricane activity will increase as the oceans warm. In the 1970s, there was an average of nine hurricanes per year. In the 1990s, we can expect at least twice as many,

253

some of them more vigorous than anything experienced in recent decades.

• By the end of the 1990s, summer drought and heat will make it difficult to grow traditional crops using conventional techniques in England and Wales, and much of continental Europe. Dust-bowl conditions as severe as those of the 1930s may return to the US. These problems may be alleviated by extensive irrigation, if water is available.

• In North America and Europe, wheat yields will decline by a quarter within twenty years. Scotland and Canada, however, will benefit from the changes, as may parts of the Soviet Union.

• Sea level will rise by at least a quarter of a metre over the next twenty years.

• And all that is just the beginning, a hint of the more sweeping changes to come in the first half of the twenty-first century.

Just how you react to all of this depends on who you are and where you live. The owner of a block of holiday apartments in southern Spain, or seafront property in Florida, will respond very differently to a farmer in Scotland or on the Canadian prairie. But it is important that you do react, if only by changing to low-wattage light bulbs, turning down the central heating, and writing to your political representative. We got ourselves, and the world, into this mess, and there is nobody else around to get us out of it. But if you are sanguine enough to take the philosophical view, and not too human-chauvinist to acknowledge the role of other forms of life in maintaining Gaia, there is a crumb of comfort to digest. Even if the greenhouse effect proceeds so fast, and so far, that human society collapses under the strain, there will still be life on Earth, and conditions may settle down to something resembling the climatically lopsided but biologically flourishing planet of five to ten million years ago. That might well be better, for every other species and for Gaia, than a world ravaged by human carelessness.

CFCs, Ozone and the Greenhouse

The ozone layer spans most of the stratosphere, roughly ten to fifty kilometres above our heads. The concentration of ozone is greatest in the region of the stratosphere from about fifteen to thirty kilometres altitude, and it is spread around the entire world. But there are so few molecules of ozone in this tenuous layer of the atmosphere that if they were all brought down to the surface and spread evenly around the globe the pressure of the atmosphere above would squeeze them into a layer just three millimetres thick. These few molecules of ozone protect us from the harmful ultraviolet rays of the Sun, and absorb the energy that keeps the stratosphere warmer than the troposphere below, acting as a lid on convective weather systems.

Ozone in the stratosphere is constantly being created and constantly being destroyed by chemical reactions that involve sunlight (photochemical reactions). Over millions of years, a natural balance had been established between the two processes, so that the amount of ozone in the stratosphere stayed roughly constant. But human activities, and in particular the release of chlorofluorocarbons, or CFCs, have increased the rate at which ozone is destroyed without any compensating increase in the rate at which it is produced. As a result, the amount of ozone in the stratosphere is now declining over the world as a whole, not just in the much-publicized 'hole' over Antarctica.*

CFCs are only manufactured by human industrial activities, and have no counterparts in nature. They have been widely used

*For more details on ozone cycles and the Antarctic hole, see my book *The Hole in the Sky* (Bantam/Corgi, 1988).

as propellants in spray cans, to blow bubbles in plastic foam, as the working fluid in refrigerators and air conditioners, and in cleaning electric circuits and microchips. They have very long lifetimes – more than a hundred years – but eventually break down in the stratosphere, releasing chlorine atoms which can attack ozone. Because of the long lifetime of CFCs, their concentration in the atmosphere will continue to increase unless the amount being released each year is cut back to *less* than fifteen per cent of the amount that was produced in 1988.

In the 1980s, attention was focused on the damage being done by CFCs by the discovery of a massive depletion of ozone from the stratosphere above Antarctica each spring. A major international research effort, which included observations from the ground, from satellites, from instruments carried by balloons and from an aircraft flying through the Antarctic stratosphere nineteen kilometres high, proved beyond doubt that this is caused by the presence of chlorine from CFCs. Over most of the atmosphere, for most of the time, the chlorine is locked up in compounds such as chlorine nitrate, and cannot immediately attack ozone. But in the cold and the dark of the stratosphere over Antarctica in winter, chemical reactions that take place on the surface of tiny ice particles in the polar stratospheric clouds lock up the 'nitrate' in the form of frozen nitric acid, and release chlorine. When the Sun returns in the spring, it triggers photochemical reactions that destroy ozone and involve chlorine monoxide. As the high-flying ER-2 research aircraft travelled south into the region of ozone depletion in 1987 (*Figure A.1*), its instruments recorded a drop in ozone concentrations that was precisely mirrored by a rise in chlorine monoxide (*Figure A.2*). This was the 'smoking gun' that firmly established the culprit responsible for the hole.

Such a depletion of ozone had not been predicted because chemists had not attempted to calculate (or mimic in laboratory tests) the effects of relevant chemical reactions involving a mixture of solid particles and gases – so-called 'heterogeneous' reactions. Their forecasts of the likely impact of CFCs on stratospheric ozone had been based on calculations for mixtures of gases alone – 'homogeneous' reactions. In the wake of the discovery of the Antarctic ozone hole, those calculations were revised and new laboratory tests were carried out. At the same time, measurements of the ozone concentration of the stratosphere around the world were re-analysed. Both lines of research suggest that there may be worse damage to the ozone layer still to come.

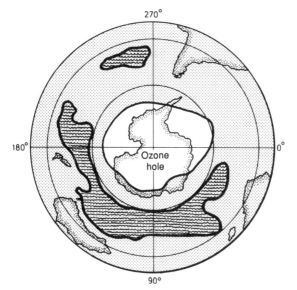

Figure A.1
The hole in the ozone layer over Antarctica in the spring of 1987. The shaded area partially surrounding the hole is the region of descending air from lower latitudes that later pushed the hole off Antarctica as it broke up in summer. This happens each year; the 'new' air is relatively rich in ozone, but also carries in chlorine compounds that interact with cloud particles in winter, releasing active chlorine that destroys the ozone the following spring and creates a new hole. (Adapted from NASA satellite data)

The particular set of heterogeneous reactions responsible for the hole over Antarctica can only occur there because temperatures fall below −80°C. Nowhere else in the stratosphere is that cold. But a slightly different set of heterogeneous reactions taking place at temperatures of 'only' about −72°C could occur in stratospheric clouds above the Arctic. There is already evidence that stratospheric ozone over the Arctic is also being depleted. As yet the effect is smaller than over Antarctica, partly because the Arctic stratosphere is warmer. But as the amount of CFCs in the atmosphere continues to increase it may reach a critical level as seems to have happened over Antarctica already, triggering the same kind of massive ozone depletion at high northern latitudes that has already been seen in the southern hemisphere. Also, as the greenhouse effect traps heat near the surface of the ground, in the troposphere, the stratosphere will cool by as much as 10°C over the next fifty years, and this will encourage the kind of heterogeneous reactions that now occur only over Antarctica to develop above the Arctic.

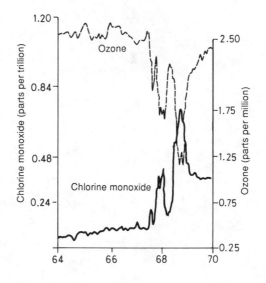

Figure A.2
On a flight into the ozone hole (Figure A.1) on 16 September 1987, at an altitude of just under twenty kilometres, instruments on board NASA's ER-2 research aircraft measured changes in the concentration of chlorine monoxide and ozone in the stratosphere. The two plots mirror each other almost exactly as the aircraft crosses the edge of the hole. Ozone decreases where chlorine monoxide increases, showing that it is chlorine monoxide that attacks ozone. The chlorine can only come from the breakdown of chlorofluorocarbons; this plot became known as the 'smoking gun' that proved CFCs are responsible for the ozone hole. (Source: James Anderson/NASA)

As if that were not enough to worry about, researchers are also concerned now about the possible impact of major volcanic eruptions on the ozone layer in the near future. In the past, eruptions such as that of El Chichón in South America in 1982 have caused a decrease in the amount of ozone in the stratosphere. This is a natural process from which the stratosphere recovers in a year or two. But with an increasing amount of chlorine nitrate in the stratosphere from the breakdown of CFCs it seems likely that the particles of dust and droplets of sulphuric acid thrown into the stratosphere by any future eruption of this kind might become involved in heterogeneous reactions that release chlorine and cause far more destruction of ozone than any past volcanic eruption has caused. The springtime hole over Antarctica is roughly as big as the continental United States and as deep as Mount Everest is tall; sometime in the 1990s, a large, explosive volcanic eruption might create a hole in the ozone layer just as big, drifting over inhabited regions of the globe and allowing solar ultraviolet to penetrate to the ground, damaging crops and causing severe sunburn in people.

The Antarctic hole itself is already doing something like this, in a more modest way. As the southern summer develops each year, the hole (the region of air depleted in ozone) is physically pushed away from the Antarctic continent by 'new' air descending into the Antarctic stratosphere from higher altitudes and lower latitudes.* The hole falls apart by early November, breaking into blobs of ozone-depleted air that wander around the southern hemisphere before filling in and disappearing. In December 1987, one of these blobs passed over a swathe of the southern hemisphere from Perth in Australia to the south island of New Zealand. Over Melbourne, a city of three million people, ozone levels decreased by almost twelve per cent for three days, from 11 to 13 December. According to calculations by the US Environmental Protection Agency, a *permanent* ten per cent increase in the amount of solar ultraviolet radiation reaching the ground could cause a ten per cent increase in the number of deaths from a form of skin cancer known as malignant melanoma, as well as causing a fifty per cent increase in the incidence of non-lethal but disfiguring skin cancers.

As the ozone-depleted air breaks up and mixes in with the rest of the southern stratosphere each summer, the overall effect should be to dilute the ozone concentration of the stratosphere above the southern hemisphere by about one per cent, compared with the concentration that existed in the 1970s before the hole

---

*This 'new' air carries with it the seeds of the next year's hole, in the form of chlorine compounds ready to be triggered into active, ozone-eating life by the cold and dark of the Antarctic winter. Although the size of the hole varies from year to year, this pattern makes it possible to forecast the severity of the ozone depletion, to some extent, about nine months in advance. If the hole is relatively small and breaks up quickly one year, it allows more chlorine-laden air to become established over Antarctica in summer, causing a bigger and longer-lasting hole the next spring. That in turn holds the 'new' chlorine-laden air at bay for part of the next summer, ensuring that the following year the hole is smaller again. There is a rough two-year cycle; but don't take this as definitive, since changes in the circulation of the southern hemisphere, linked with the ENSO phenomenon, can shake up the pattern and produce two years in a row with either a relatively weak hole or a relatively strong hole. It is worth noting, by the way, that the depletion of ozone over Antarctica in spring is now so severe that when the experts talk of a 'weak' hole they mean that 'only' twenty-five per cent of the ozone in the stratosphere is destroyed. Before 1985, any scientist who had suggested that as *much* ozone as this could disappear from the Antarctic stratosphere in a few weeks each spring would have been dismissed as crazy.

began to appear. In fact, the amount of ozone in the southern stratosphere seems to have declined by six per cent over the past twenty years but this only came to light when the discovery of the Antarctic ozone hole made researchers go back to old records and analyse them in a new way.

Earlier studies had, mistakenly, combined all of the data to give global annual averages for measurements of the amount of ozone in the stratosphere. This, it turns out, masks changes that occur in high latitudes in winter. The measurements are made from the ground, using instruments known as Dobson spectrophotometers, that analyse the spectrum of light (from the Sun) that has passed through the ozone layer. If there is more ozone, this shows up in the form of stronger spectral lines; if there is less ozone, the lines in the spectrum are weaker. The best data comes from the northern hemisphere, where most people live and where there is a good network of Dobson stations. They show that between 1969 and 1986 there was a decline in wintertime ozone concentration of the stratosphere between 53°N and 64°N of about six per cent. Between 40°N and 52°N the decline is just under five per cent; and between 30°N and 39°N there are hints of a smaller, but not statistically significant, decline. Less complete measurements for the southern hemisphere suggest that the pattern is similar there, and that the overall decline in the ozone concentration of the stratosphere was about 2.5 per cent.*

The main band of northern latitudes that is affected, from 30°N to 64°N, covers the whole of the United States and runs from the southern shores of the Mediterranean to above the northernmost tip of Scotland (*Figure A.3*). There is every reason to expect that ozone has been depleted at still higher latitudes, but the measurements to prove this are not available. In the worst case so far recorded, ozone over a latitude band that includes Dublin, Moscow and Anchorage in Alaska was eight per cent less in January 1986 than it had been in January 1969. Theories based on homogeneous chemistry had suggested that the decline in ozone concentration due to the build-up of CFCs over that period should be no more than two per cent. The strong suspicion among the experts is that

*The ground-based measurements are backed up by satellite observations. These give global coverage, but have to be calibrated by comparison with measurements from ground stations that they pass over. Ironically, with the advent of satellites some ground-based stations have been closed in a cost-cutting exercise; there is now an urgent need for a new ground-based Dobson network to monitor the continuing decline in ozone, *and* for new satellites.

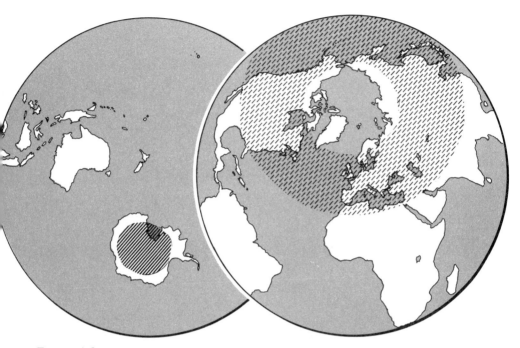

Figure A.3
The region of the northern hemisphere where measurements show a wintertime depletion of ozone over the past twenty years, and the region over Antarctica affected by the notorious 'hole in the sky'.

heterogeneous processes are already at work in other parts of the globe, outside the polar regions.

All of this has important implications for human activity, and will be the subject of a major research effort in the 1990s. Tests show, for example, that crop species such as maize, wheat, soya beans and rice produce lower yields and lower quality if they are subjected to increased amounts of ultraviolet radiation (nobody has tested their response to a simultaneous increase in ultraviolet, carbon dioxide and heat, with a decrease in water; that, however, is the likely prospect for crops in the US Midwest in the coming decades). But in the context of the greenhouse effect, there is another important connection. Just as the greenhouse effect, by cooling the stratosphere, increases the destruction of ozone, so the destruction of ozone, by allowing ultraviolet radiation to penetrate to the surface, may strengthen the greenhouse effect. There is certainly a direct strengthening of the greenhouse warming of the globe, as energy that would otherwise be absorbed in the stratosphere penetrates deeper into the atmosphere. But another process may make things far worse.

The main region of the globe affected by ozone depletion so far is the Antarctic. Fortunately, the land mass of Antarctica is largely uninhabited, and there is no evidence yet that either people or penguins have been adversely affected by the increased dose of ultraviolet they have received in recent years. But the seas around Antarctica are teeming with life, and as we have seen in this book the phytoplankton that live at high latitudes play a key role in the natural carbon cycles of Gaia, and probably in feedbacks involving the astronomical rhythms of ice ages. Research shows that phytoplankton exposed to increased levels of ultraviolet radiation are less productive and fix less carbon from the atmosphere. John Calkins, of the University of Kentucky, has carried out tests in which plankton in tanks of water were exposed to ten per cent more ultraviolet than normal. Mobile plankton retreated to darker regions of the tank in response. There they seemed to come to no harm, but because they had moved to darker water they received less light overall, not just less ultraviolet. Calkins calculates that at the depths that cancel out the brighter ultraviolet the plankton are losing up to five per cent of the light they need for photosynthesis. If similar changes occur in the oceans as stratospheric ozone is depleted and more solar ultraviolet penetrates to the surface of the sea, plankton will become less efficient at fixing carbon from carbon dioxide in the air through photosynthesis. The build-up of carbon dioxide in the atmosphere will then proceed even more rapidly than expected, increasing the strength of the greenhouse effect above the range of forecasts discussed in this book.

This is in addition to problems that may be caused by increased ultraviolet to fish larvae in the oceans (a ten per cent increase in ultraviolet could kill *all* fish larvae in the top ten metres of the sea), to crops and animals on land, or to human beings. And, don't forget, the CFCs that cause the destruction of ozone in the stratosphere are also major direct contributors to the greenhouse effect itself. In view of all this, it may come as an unpleasant shock to be told that the widely publicized action of many governments in signing the Montreal Protocol to limit release of CFCs is not what it is sometimes cracked up to be. That agreement calls for a freeze of consumption of CFCs by 1990 followed by a twenty per cent cut in 1994 and a further reduction of thirty per cent of the base figure by 1999. It sounds good. But CFCs have such a long lifetime in the atmosphere that at present only one sixth of the amount released each year is broken down and taken out of circulation (that is, turned into a form that has the potential to destroy ozone) by natural processes.* Simply in order to maintain the present concentration of CFCs, which has proved so damaging over Antarctica and is already exerting an influence around the world, would require an immediate cut back to fifteen per cent (that is, a cut of eighty-five per cent) in the production and release of CFCs. The only hope of *reducing* the impact of CFCs on the ozone layer and the greenhouse effect is to have an immediate and total ban on their production. Even then, there would still be five times as much CFC floating around in the atmosphere, waiting to be converted into an ozone-eating form, as the amount that is actually broken down each year and has released its chlorine.

Countries such as India, China and Brazil are not even signatories to the Montreal agreement, and they have burgeoning CFC-based industries, largely in refrigeration. The Montreal agreement, even if fully implemented, commits us to an increase in stratospheric chlorine to three times 1988 levels by 2020. Nobody knows what new ozone-eating reactions may be triggered as a result. Even if direct release of CFCs stopped immediately, ozone depletion would continue to increase for two decades as the existing atmospheric reservoir of CFC filtered up into the stratosphere. About forty per cent of the CFC molecules now in the atmosphere will still be there in 2100, even if no more are released. But at least the prospect of bringing a halt to the release of CFCs, and (eventually) allowing the ozone cycles of Gaia to return to normal is a real and feasible one, if somewhat remote. There is no prospect at all of bringing a halt to the release of carbon dioxide and other greenhouse gases, and thereby allowing the carbon cycles and the temperature of Gaia to return to normal.

*But the amount broken down each year will be less if the concentration of CFCs falls (the sinks become less efficient), so that it would take far longer than six years to clear the air of these gases even if no more were being released. In fact, after a hundred years only a little over half of the CFC would be gone.

Bibliography

A great deal of the information presented in this book came from personal discussions with researchers involved in the greenhouse debate, and from their scientific papers published in the research journals or presented at major conferences on the subject. The best way to keep up to date on new developments is through the news pages of the journals *Science* and *Nature*, or at a more accessible level from the magazine *New Scientist*. The books listed here provide useful background information on topics mentioned in the text; the ones marked with an asterisk are really for specialists, but the rest are accessible for anyone with an interest in climatic change.

*Bert Bolin, Bo Döös, Jill Jäger and Richard Warrick (editors), *The Greenhouse Effect, Climatic Change, and Ecosystems*, Wiley (Chichester), 1986.
The most complete and authoritative single-volume guide to the greenhouse effect, prepared by a team of experts for the Scientific Committee on Problems of the Environment (SCOPE). Reasonably accessible for anyone with a serious interest in environmental issues (and essential reading for all politicians), but by no means a light read – 541 pages of closely argued discussion.

Georg Breuer, *Air in Danger*, Cambridge University Press (Cambridge), 1980.
Slightly out of date, but a sound basic guide to the nature of the Earth's atmosphere and fundamentals of the greenhouse effect.

Hermann Flohn, *Possible Climatic Consequences of a Man-Made Global Warming*, IIASA (Laxenburg), 1980.

Anonymous, *Life on a Warmer Earth*, IIASA (Laxenburg), 1981.

Hermann Flohn's study of the kind of climate we might experience before the end of the twenty-first century is based on studies of past climates, including the post-glacial optimum and the previous interglacial. The short (81-page) booklet is itself clear and intelligible to anyone prepared to make a tiny effort to come to grips with a few technical terms. The International Institute for Applied Systems Analysis, for which Flohn carried out the study, decided, however, to produce an even more easily readable version for the general public, called an 'Executive Summary' and boiled down to 59 pages; this must rate as the best instant guide to the magnitude of the changes we are likely to witness in our lifetimes. Unfortunately, both versions may be difficult to lay hands on. Originally, single copies of the 'Executive Summary' were available free of charge from IIASA, A-2361 Laxenburg, Austria – but that was in 1981. IIASA publications are also sold by NTIS, 5285 Port Royal Road, Springfield, VA 22161, USA.

John Gribbin, *Genesis*, Delta (New York) and Oxford University Press (Oxford), 1981.

A look at the origin and evolution of life against the background of events in the physical world.

John Gribbin, *The Hole in the Sky*, Bantam, Corgi (New York/London), 1988.

As well as contributing to the greenhouse effect, chlorofluorocarbons destroy ozone in the stratosphere. This book tells about the discovery and investigation of the hole in the ozone layer over Antarctica.

John Gribbin and Mick Kelly, *Winds of Change*, Headway (London), 1989.

Hothouse Earth deals primarily with the scientific basis of the greenhouse effect, and the scenarios for the hothouse world of the twenty-first century. *Winds of Change* takes most of these scenarios at face value, and, drawing on Kelly's work with the Climatic Research Unit at the University of East Anglia, looks more closely at the social and political implications of the changes we face, and at the action needed to alleviate the problems that will arise.

John Imbrie and Katherine Palmer Imbrie, *Ice Ages: Solving the Mystery*, Enslow (New York), 1978.
A readable and comprehensive historical account of the astronomical theory of ice ages (the Milankovitch Model) up to the point where John Imbrie, in 1976, was co-author of the paper that proved the connection between the astronomical cycles and ice age rhythms. Nothing here, though, on the 1980s work which linked the cycles with changes in the carbon dioxide greenhouse effect.

William Kellogg and Robert Schware, *Climate Change and Society*, Westview Press (Boulder), 1981.
An excellent overview of the problems faced by society in responding to the challenge of climatic change, and especially to the greenhouse effect. A good overview for the casual reader, but full of references to follow up if you have a more serious interest.

Emmanuel Le Roy Ladurie, *Times of Feast, Times of Famine*, Doubleday (New York), 1971.
A historian's view of the climatic changes of the past millennium or so. Interesting insights into natural climate fluctuations, but only marginally relevant to how the world is likely to warm in the twenty-first century.

*H. H. Lamb, *Climate: Present, Past and Future*, Methuen (London), Volume 1, 1972; Volume 2, 1977.
These two volumes provide a summary of a life's work studying variations in climate and their implications. Volume 1 explains how climate works, while Volume 2 looks at climates of the past, including the rhythm of ice ages and the fluctuations since the latest ice age. Books for the connoisseur, and priced accordingly, but full of interesting information.

James Lovelock, *Gaia*, Oxford University Press (Oxford), 1979.
Lovelock's first, landmark book on the Earth as a living organism.

James Lovelock, *The Ages of Gaia*, Norton (New York) and Oxford University Press (Oxford), 1988.
Lovelock's second detailed look at Gaia, taking on board and responding to criticism of the original hypothesis (now grown up into a full-blown theory). Both are beautifully written, thoughtful and, in the best sense of the word, provocative. This one is the

266

better book, but both are well worth reading.

Amory Lovins, *Soft Energy Paths: Towards a Durable Peace*, Friends of the Earth, Ballinger (Cambridge, Massachusetts), 1977.
Written in response to the energy crisis caused by oil price rises in the early 1970s to show how the world could sustain a high standard of living without depending on either fossil or nuclear fuels. Idealistic and over-optimistic in so far as the full 'soft energy' option is not likely to be realized in our lifetimes, but an invaluable guide even today to many of the kinds of energy sources that could, if taken up by governments, go some way towards reducing the rate at which greenhouse gases are building up in the atmosphere.

Eugene Mallove, *The Quickening Universe*, St Martin's Press (New York), 1987.
A somewhat 'hippy' view of the relationship between life and the Universe, easy to read and even easier to forget, which includes a chapter on Gaia.

*Margaret Mead and William Kellogg (editors), *The Atmosphere: Endangered and Endangering*, Castle House (Tunbridge Wells), 1980.
The proceedings of a conference, partly inspired by the concept of Gaia, about the interaction between mankind and the atmospheric environment. Much of the content has been overtaken by events, and it is too technical for the casual reader, but it has considerable historical interest for anyone with a serious concern for these problems.

*Martin Parry, Timothy Carter and Nicolaas Konijn (editors), *The Impact of Climatic Variations on Agriculture* (two volumes), Kluwer (Dordrecht), 1988.
The definitive (and only) comprehensive study of how the greenhouse effect is likely to affect agricultural activity around the globe in the hothouse world of the twenty-first century. Strictly for those with a serious interest in the problem – but if you *are* a politician or planner, or a farmer, this is essential reading.

*Graeme Pearman (editor), *Greenhouse: Planning for Climate Change*, E. J. Brill/CSIRO (Leiden), 1988.
A compilation of the scientific papers presented at a conference on the greenhouse effect held in Melbourne late in 1987. Really for the specialists, but particularly important because it provides a

perspective on the problem as it affects the southern hemisphere. Most research reflects the fact that most researchers live in the north!

Stephen Schneider and Randi Londer, *The Coevolution of Climate & Life*, Sierra Club Books (San Francisco), 1984.
A massive (563-page) labour of love from climatologist Schneider and writer Londer. One of the most complete and accessible single-volume guides to the whole business of climatic change, especially interesting because of its exposition of the concept of 'coevolution'. This sees life and the environment evolving together and influencing each other, but is distinct from the idea of Gaia, which, at least in its original form, has life dominating the global system.

Jonathan Weiner, *Planet Earth*, Bantam (New York), 1986.
The companion volume to an acclaimed seven-part TV series of the same name. Very easy to read, beautifully illustrated, it will add little to what you already know about the greenhouse effect, but is informative on how Jim Lovelock developed the idea of Gaia.

Index

269

271